MAINTENANCE OF
PROCESS PLANT

a guide to safe practice

Secon

Arthu

INSTIT

The information in this guide is given in good
faith and belief in its accuracy, but does not
imply the acceptance of any legal liability or
responsibility whatsoever, by the Institution, the
editor, or any individual member of the Working
Party for the consequences of its use or misuse
in any particular circumstances.

Published by
Institution of Chemical Engineers,
Davis Building,
165–171 Railway Terrace,
Rugby, Warwickshire CV21 3HQ, UK

Copyright © 1980 and 1992
Institution of Chemical Engineers

ISBN 0 85295 292 9

First Edition 1980
Second Edition 1992

Printed in England by BPCC Wheatons Ltd, Exeter.

16/11/92.

PREFACE TO THE SECOND EDITION

The preparation of the First Edition of this guide was begun some years before its publication in 1980 and, as a considerable amount of time had elapsed since then, it was felt that it would be worth undertaking a fairly simple update.

Upon starting the revision, however, it soon became apparent that the valuable guide produced by the original Working Party could be substantially improved upon, not only by updating the subject matter where it was obviously necessary, but also by altering the balance and emphasis of the book. As a result, the book came to be virtually rewritten rather than simply revised, as was the original intention.

The First Edition of the book contained most of the material relevant to its purpose, but certain matters of a peripheral nature were accorded, it was felt, more space than was perhaps justified. These included such subjects as scaffolding, ladders, lifting tackle and electrical apparatus which, whilst certainly of vital importance in safe plant maintenance, were not felt to be in the main stream of interest to chemical engineers.

Thus, in the Second Edition the whole balance of the book has been altered; it is now written principally for chemical engineers, with emphasis more on the 'chemical' hazards which are likely to be encountered during maintenance work, such as entry into confined spaces, contact with toxic and flammable substances, etc. The other 'non-chemical' risks referred to above have not been excluded; information on them has been brought up to date whilst the treatment of them in the text has been reduced and replaced by comprehensive reference sources for further reading.

All the references to legislation, standards and other documents throughout the text have been updated and much new material has been introduced for the first time.

The Second Edition of the book has a new title which is intended to cover its new subject matter more satisfactorily than would that of the First Edition. It is hoped that it will draw the attention of the potential reader to the whole subject of safer plant maintenance, from the preparation for the work, to its implementation and to its final completion, and further, that it will demonstrate how such work can be carried out without danger to those engaged upon it. If this can be accomplished in a real-life practical working environment, then the guide will have fulfilled its purpose.

Arthur Townsend has had a lifetime's experience in practical process safety in the chemical and allied industries, much of it in a senior position in the Health and Safety Executive. He has used this experience to produce a publication which will be of value to all who are engaged upon, or associated with, the maintenance of process plant and are committed to carrying this out safely.

PREFACE TO THE FIRST EDITION

This guide has been prepared by a Working Party of the Engineering Practices Committee of the Institution of Chemical Engineers. Its objective is to provide a brief working basis for all concerned in the preparation of chemical process plant for maintenance.

Larger organizations operating plant of this kind will already have excellent and well-established procedures for this purpose and, while this guide may therefore be of greater assistance to the small organization, it is intended as a wider work of reference. Recently graduated engineers, of most disciplines associated with chemical and process plant, should also find this guide helpful.

While providing a basis for good practice in most situations, there will be situations where strict adherence to the stipulations in this guide will not be appropriate or where factors not mentioned in it may apply.

The composition of the Working Party was:

B.E.A. Thomas(Chairman)	Chemical and Thermal Engineering Limited
J.M. Connolly	Organics Division, ICI Limited
M.J. Hagger	British Celanese Limited
J. Hallam	Dista Products Limited
J.E. Hickford	Carless Colvents Limited
S.P. Kavanagh	Shell Chemicals UK Limited
J.F. Thorley	Dista Products Limted
E.F.D. Winter	Glaxo Holdings Limited

Acknowledgements are due to:

H.A. Anson, May & Baker Limited, who carried out preliminary work before the Working Party was formed, and continued as a corresponding member.

T. Kletz, Industrial Professor of Loughborough University, for his help and providing information in this guide.

H.J. Foxcroft, Double L Consultants Limited for some of the contributions on planning and organization of maintenance.

CAPITB, for permission to reproduce summaries incorporated in the appendices on protective clothing, statutory examination intervals and scaffolding.

The following provided information on relevant company procedures and permit documents:

British Celanese Limited
BP Chemicals Limited
Dista Products Limited
Dow Chemical Company Limited
Glaxo Holdings Limited
ICI Limited
Monsanto Chemicals Limited
Rohm & Haas Limited
Shell Chemicals UK Limited

The Fire Research Station provided literature references on dust explosion hazards.

Many others have collaborated in the preparation of this guide, and their help is acknowledged.

CONTENTS

1. INTRODUCTION

1.1 GENERAL

Whilst maintenance work on chemical and allied plant needs to be economically justified, it should be conducted with care for the safety of those involved in it. This guide has been produced with that principle in mind, as a basis for the development of an in-plant scheme for the proper preparation of plant for maintenance and the safety of personnel whilst carrying it out.

1.2 LEGAL AND OTHER REQUIREMENTS

There are legal requirements, including the Factories Act 1961 and the Health and Safety at Work etc Act 1974 (HSW Act), which require employers to provide both a safe place and a safe system of work. There are additional regulations, such as the Electricity at Work Regulations 1989 and the Highly Flammable Liquids and Liquefied Petroleum Gases Regulations 1972, which give more detailed legal requirements than do the principal statutes. The object of regulations such as these is to achieve and maintain a safe working environment; they will be referred to in greater detail later.

References in this booklet to Statutory Regulations, British Standards and other publications are correct at the time of publication. They are, however, being constantly revised and updated and so, if detailed guidance is required from them, reference should always be made to the current version.

1.3 PLANNING AND ORGANIZATION

The first step in preparing to carry out maintenance work should be planning and organising it. As part of this preparation, it is important to ensure that the maintenance organisation establishes clear lines of responsibility and authority, particularly regarding permit-to-work systems and procedures for authorizing and recording plant modifications.

Proper preparation for maintenance is particularly important in the chemical and similar industries, since many of the raw materials, intermediates and products which are handled can be hazardous. They may be flammable, explosive, toxic, or corrosive; they may be hot or under pressure. Thus, wherever

1

possible, all hazardous materials should be removed before equipment is opened up for inspection, repair or modification. If complete removal is not possible, then the operatives carrying out the work must be told of the hazards which are present and the safety precautions to be taken.

1.4 SAFE SYSTEMS OF WORK

Work should never be carried out in a potentially hazardous area without a proper permit-to-work system. It should be emphasized to all parties concerned that the system is not intended to make life difficult for them, but that its purpose is to ensure their personal safety.

The system should ensure that those persons doing the work and others who may be associated with them are not exposed to danger, that the work does not cause danger to others not directly concerned with it, and that there is compliance with legal requirements, technical standards and codes of safe practice.

Permits-to-work or clearance certificates are required for work in the following types of situation:

- confined spaces;
- locations where accidental or unauthorized starting of plant or equipment may endanger others;
- on conveyors, lifts, hoists and cranes;
- where toxic, flammable or corrosive substances are present;
- where lack of oxygen or, just as important, oxygen enrichment can occur;
- where hot-work is to be done on plant or equipment which has, at any time, contained chemical substances of any kind;
- in any untypical circumstances which are not covered by routine precautions or instructions.

The following principles should form the basis of a permit-to-work system:

- Isolation of equipment must be complete and, once achieved, must remain secure.
- Any residual hazard must be properly dealt with.
- All equipment and plant locations must be correctly and clearly identified.
- Maintenance workers must be adequately trained and instructed.
- Unauthorized alterations to the work programme must not be permitted.

- The system must be properly monitored.

The preparation of a permit-to-work certificate involves two groups of people:

- the maintenance team (normally the acceptors);
- the production team (the operators of the plant).

The persons responsible for issuing permits-to-work may be drawn from both these groups and will be appointed by virtue of their training, experience on the plant and other relevant qualifications, rather than because of any position held by them in the organization.

1.5 CONTRACT LABOUR

Special consideration must be given to potentially hazardous operations if a contractor's labour force is present. These workers must be made aware of hazards on the plant where they are working, and similarly on adjoining plants; they must be familiar with all relevant safety and certification procedures.

1.6 TRAINING

Everyone involved needs training in operating the system. But, even more important, they should be convinced of the need for it and that its principles should be followed at all times. There must be a validation process to ensure that the training has met its objectives.

During training it should be stressed that many accidents have occurred in factories allegedly working to a 'permit' system because either the system was not satisfactorily designed or it was not properly used. Incorrect preparation and handback of equipment during maintenance work is a common cause of serious accidents in the chemical and allied industries.

1.7 MONITORING

Once the permit-to-work system has been set up and the personnel trained in its use, it must be constantly monitored. If accidents are to be avoided, this must be done regularly and immediate action taken if any irregularities are found. The following matters should form the basis of the monitoring system:

- spot checks on the use of the permit, on the plant;
- questions and tests to establish the competence of the authorized persons;
- analysis of the job card against the corresponding permit;

3

- immediate detailed investigation of any accident which occurs under permit-to-work conditions;
- joint consultations between management and workers about the effectiveness of the system.

2.   ORGANIZATION AND PLANNING OF MAINTENANCE

2.1   GENERAL

The chemical and allied industries are capital intensive and must maximize plant and equipment operating time, and so the planning and organization of effective maintenance shutdown periods is important. The actual procedures and their benefits depend on the size and structure of the organization. This section is intended to be used as it relates to individual situations, as appropriate.

Maintenance plans will only be successful if each person, no matter what his position, in whatever size of organization, understands his role and is accountable for its fulfilment.

Many engineering jobs which arise on a plant are of less than a couple of hours' duration, involve only one trade and use a minimum of materials. These small jobs may amount to as much as 80% of all maintenance jobs. To deal with these small tasks, the process plant in a factory is often grouped into production units, each of which includes in its production team sufficient engineering expertise to cope with most of them. The aim should be that each production unit deals with its own short term maintenance problems under the philosophy 'a stitch in time saves nine'. The expertise in the production unit would be in the form of process men who are trained to carry out certain types of engineering work and to assist skilled tradesmen who are permanently attached to the unit.

2.2   PLANNING

Planning may be defined as the ordering of events to ensure the economical use of resources, so as to achieve a defined objective. Smaller jobs, as well as the bigger and more complex ones, can benefit considerably from a disciplined, pre-arranged approach to planning.

In considering a formal procedure which will enable the maintenance support groups to deal with the larger tasks, a procedural chain of actions through which each job will pass is required, as follows:

• Initiate job request.
• Classify job and assign priority.
• Specify details of the work to be done.

5

- Estimate the time required and materials costs.
- Obtain approval to proceed (particularly important in the case of plant modifications).
- Order, chase and obtain materials and equipment.
- Schedule resources, ie arrange when job is to be done.
- Carry out the work.
- Follow hand-back procedure on completion.
- Organize any feed-back, eg uncompleted jobs, extra work required.
- Prepare proper records of the work done (eg update P&I diagrams and other documents), paying particular attention to modifications.

When a plant is to be shut down for maintenance, the work should be planned as follows:

2.2.1 WORK CONTENT OF THE PLAN

The plan should cover shutdown procedures, plant preparation, hand-over to engineering personnel, the work to be done (including scheduled shutdown preventative maintenance and statutory inspections), hand-back to production and recommissioning. Lack of attention to the beginning and the end often causes unnecessary expense, wasted effort and over-run of the shutdown time-table. It is important, therefore, that both engineering and production staff are involved in agreeing what work is to be done. Items which cannot be overlooked will often arise as work proceeds but, if the plan is to be realistic, sufficient pre-planning must be carried out to keep these to a minimum.

The content of the plan might be based, for example, on information obtained before the shutdown by:

- taking advantage of short plant closures (planned or unplanned) to carry out a pre-shutdown examination to find out what work may be required;
- monitoring the condition of the equipment using vibration monitoring, ultrasonic thickness testing, pressure drop measurements, temperature profiles, etc;
- keeping a history of plant performance so that failure of particular components can be anticipated.

2.2.2 TIMING

The precise date of a planned shutdown will often be decided by the need for statutory inspections. These cannot be put back beyond the due date if the equipment is to be kept in operation. Having decided upon a start date (a yearly programme is helpful), the planning stage must precede this by a sufficient period of time to organize the availability of spares, manning and support services (eg rigging, scaffolding, cranes and specialist contractors).

2.2.3 PREPARING THE PLAN
There are several ways in which this may be done:

- bar charts — these have a fixed time scale;
- scheduling — this has the advantage that there is freedom to adjust the plan;
- critical path analysis;
- computer techniques;

Each method has its advantages and disadvantages and the one chosen may depend on the work to be planned, what is wanted from the plan, and the planning resources available.

Each group or trade to be involved in the shutdown must be relied upon to fulfil its role and this can only be done if each has a say in the creation of the plan, so that it can set its own individual targets.

2.2.4 IMPLEMENTING THE PLAN
The team carrying out the work must understand the plan and be committed to its objectives. The plan must be realistic: it should be based on individual working methods, and on the time and organisation which they entail. The following points are helpful in implementing planned maintenance shutdowns:

- A co-ordinator, usually from the engineering department, should be appointed.
- A shutdown meeting, which should include representatives of both the engineering and production teams, should be held to introduce and explain the plan.
- A copy of the plan should be displayed on or near the working area.
- Progress should be marked on a copy of the plan as the work proceeds, based on reports from supervisors at the end of each shift or day.
- The plan should be revised before each new work period begins and copies of the updated plan for the next day be sent to the production management so that they can have permits-to-work ready in advance.
- Post shutdown meetings should be held, involving both engineering and production teams to discuss performance during the work and any problems encountered. The performance should be analysed and compared with the original plan. Record lessons learnt; do not re-invent the wheel at each annual shutdown!

A daily meeting between the maintenance supervisors and the planning team will build up a team approach to getting the work done. The supervisor must agree with the plan before he accepts it, and should be encouraged to make any last minute changes which he thinks are necessary, rather than to set out with a plan of work which he cannot meet. It must be established by trust that the group is being set a reasonable target.

The maintenance supervisor should regard the planning operation as a support function designed as a service to him. He should be confident that, having accepted the plan for the next day, any materials which he will require will be to hand, permits-to-work will be ready and he will not be kept waiting for other tradesmen and contractors. He can thus concentrate on the current day's work, leaving the planners to organise the following day.

Production supervisors should be encouraged to visit the planning office and to take part in the planning operation, and especially to give guidance on job priorities.

2.2.5 MODIFICATIONS AND ALTERATIONS

When a plant resumes production after a maintenance shutdown, it must be safe to operate. It is essential that all modifications, however minor, are formally approved and are designed to the proper and appropriate engineering standard. It is prudent to define a modification as *any* change to existing equipment or plant (including pipework), except when the replacement is identical in all respects but age or condition. Particular attention should be paid to temporary bypasses around equipment.

Management quite often takes the opportunity of making minor changes to process plant during a shutdown. If the alteration is essential, any possible hazards which may be introduced must be checked out, the alteration carefully considered by all qualified parties and, finally, formally approved in writing by a competent person. No modification, however small, should be exempt from this formal procedure.

Modifications may not be limited simply to physical alterations; they may also include intangibles such as operating procedures and conditions, and software associated with computerized and instrumental control systems.

When modifications which have been formally agreed have been implemented, they should be properly recorded, eg in P&I diagrams, plant operating and maintenance manuals and in any relevant engineering drawings.

3. PREPARING PLANT FOR MAINTENANCE WORK

3.1 GENERAL

Before process plant is inspected, repaired, cleaned out or modified, the following points must be considered:

- depressurizing;
- cooling down;
- removal of hazardous liquids;
- removal of solid residues;
- the need for specialized cleaning techniques;
- isolation from other plant or equipment;
- provision of safe means of access and escape for maintenance operatives;
- carrying out tests in the working area to confirm that it is safe.

These steps may not be carried out in the order shown. For example, liquids may be removed before depressurizing, or isolation may take precedence.

3.2 DEPRESSURIZING

The method to be used will depend on the hazardous nature of the material remaining in the plant.

Inert gases can be vented with care to a high-level stack or vent pipe. Flammable or toxic gases should, if possible, be vented to a lower pressure section of the plant. If this is not possible, they should pass to a flare stack or to a scrubbing system or, as a very last resort, to a safe place in the open air.

Liquids which are under elevated pressure at temperatures above their atmospheric boiling point will probably need to be cooled down to prevent excessive flashing-off of vapour when depressurized.

3.3 COOLING

During natural or accelerated cooling, it is essential that a vacuum is not created in a vessel which is not designed to withstand negative pressures. Inert gas or air should be admitted to maintain atmospheric pressure. Inert gas is preferable but, if air is used, a large excess should be admitted so as to dilute any dangerous atmosphere to a safe concentration. If the presence of a source of ignition, eg from pyrophoric residues, is suspected, then an inert gas must be employed.

Adequate time must be allowed for the cooling down of hot furnace refractories and the like.

3.4   ISOLATION OF PROCESS VESSELS AND PIPEWORK

Before attempting to remove all hazardous substances from the section of the plant which is to undergo the maintenance work, it must be physically and effectively isolated to prevent the transfer of dangerous materials into it from other units. All electrical apparatus, and any machinery which is capable of being set in motion, should be isolated in such a manner that it cannot be inadvertently switched on whilst work is being carried out in a vessel, for example.

Methods available for isolation, in ascending order of effectiveness, are: closed and locked valves; double block and bleed valve systems; spades or line blinds; and physical disconnection and blanking-off (see Figure 1).

Even when nominally shut, valves may still pass liquids or gases through them. They should not, therefore, be used as a means of isolation except

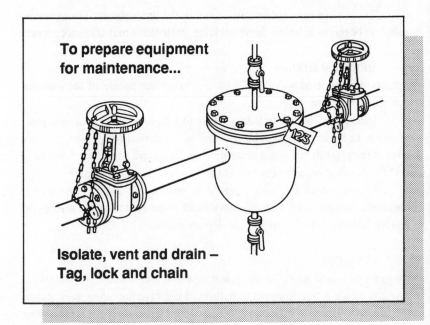

To prepare equipment
for maintenance...

Isolate, vent and drain –
Tag, lock and chain

Type A   For low risk fluids

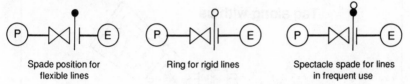

Spade position for   Ring for rigid lines   Spectacle spade for lines
flexible lines   in frequent use

Type B   For hazardous fluids with vent to check isolation

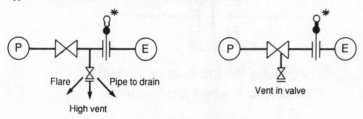

Flare   Pipe to drain   Vent in valve

High vent

Alternative destinations according to hazard

**Type C   For high pressures (>600 psi) and/or high temperatures or for
          fluids known to have isolation problems**

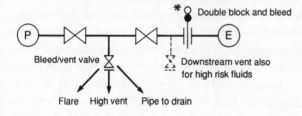

Double block and bleed

Bleed/vent valve

Downstream vent also
for high risk fluids

Flare   High vent   Pipe to drain

Type D   For steam above 600 psi

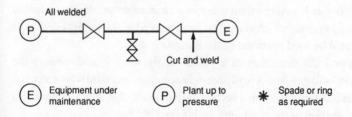

All welded

Cut and weld

(E) Equipment under   (P) Plant up to   * Spade or ring
    maintenance   pressure   as required

Figure 1   Isolation for maintenance

11

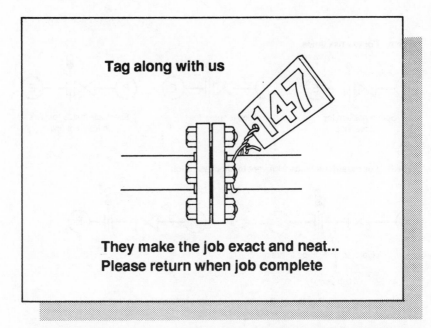

Tag along with us

They make the job exact and neat...
Please return when job complete

for hazard-free jobs which involve low risk fluids and where the isolation can be proved to be effective, eg by testing at drain points or vents.

Double block and vent systems are more effective than single valves, but do not provide adequate isolation for long periods. One of their main purposes is to enable blanking plates to be inserted safely downstream of them.

Spades, spectacle plates or proprietary line blinds should be strong enough to withstand line pressure and be of a type which, if they leak, do so to the outside of the vessel rather than inside it. The hazards which may be encountered when inserting a spade should be carefully considered before work starts. For relatively safe materials at moderate temperatures and pressures, it may be sufficient to shut and lock-off the upstream valve, then depressurise and drain the line, before breaking a flange. For more hazardous substances, a double block and bleed system, which should be incorporated in the original design of the plant, should be used upstream of the flange.

Physical disconnection of a vessel or pipework is undoubtedly the safest means of isolation, and is really the only acceptable method where a vessel is to be entered. After removing a section of pipework, that which connects to the still 'live' section of the plant must be blanked off.

Where isolation is to be carried out by any of the above methods, the appropriate valves and flanges should be identified by labels and these should be recorded on the permit-to-work.

The need to isolate sewers and drains in the vicinity of plants where maintenance work is to be done should receive careful consideration. These can provide interconnection between different plants and may thus spread a dangerous material outside the one where the maintenance shutdown is in progress.

3.5 ISOLATION OF ELECTRICAL APPARATUS AND MACHINERY

3.5.1 GENERAL

Where personnel are required to work where they may be endangered by exposure to live electrical apparatus, or to moving machinery, then these must be properly isolated from the source of power before the permit-to-work can be issued. This philosophy is embodied in the Electricity at Work Regulations 1989; detailed advice on electrical safety is given in the Memorandum of Guidance on these[1].

The safeguards and precautions which which are required to achieve safe electrical and mechanical isolation must be an integral part of the safe system of work. They are supplementary to the other forms of isolation, which are aimed chiefly at providing protection against chemical hazards. They should never be used as the primary means of protection against the latter.

Generally in preparing chemical plant for maintenance work, several methods of electrical isolation are used. The one chosen will depend on the nature of the apparatus and on the work to be done; it must remain in place for the whole time during which persons are at risk. Possible options include:

• Single locking of a suitable isolator in the OFF position. This gives acceptable isolation in most cases where mechanical work is to be carried out on items of equipment.

• Double/multiple locking of a suitable isolator in the OFF position. This gives acceptable isolation where a vessel is to be entered. Each lock must have only one key so that each person who enters can keep the key on his person whilst work is in progress.

• Withdrawal of fuses. This must be done only by electrically competent persons. It is required where suitable isolators do not exist, eg with instruments, heaters, lighting circuits, and office and laboratory equipment. A notice should be affixed to the fuse box giving notification that the fuses have been withdrawn for isolation purposes.

• Electrical disconnection at the terminal box must only be done by electrically competent persons. It may be used in conjunction with the foregoing methods to give additional security.

Before work commences, it is advisable to try to start the equipment, taking into account all possible interlocks, as a final check that effective isolation has been achieved.

NOTES
1. Jamming or locking stop/start buttons are NOT effective means of isolation.
2. All items of electrical apparatus should be permanently labelled. It must be possible to identify all parts of a circuit, eg in the case of a prime mover — the start/stop button, the isolator/conductor and the drive unit.

3.5.2   HIGH VOLTAGE SYSTEMS (TYPICALLY ABOVE 1000 VOLTS A.C.)
There are special dangers in isolating this type of system, which must be carried out by electrical personnel who have been specially trained in and authorized to use suitable control procedures. British Standard 6626 details these procedures and the precautions for dealing with high voltage systems[2]. In particular, examples are given of suitable forms which cover access to apparatus and permits-to-work. Formal procedures must be observed before any work on high voltage apparatus is undertaken.

NB Area Electricity Companies provide a maintenance service for high voltage apparatus if suitably trained personnel are not available in-house.

3.5.3   MEDIUM AND LOW VOLTAGE SYSTEMS (TYPICALLY BELOW
         1000 VOLTS A.C. AND GENERALLY 415 VOLTS A.C.3-PHASE)
Most electrical apparatus found on process plants falls into this category. There are several methods of making such apparatus safe, but it is essential that anyone involved in the isolation of medium and low voltage systems must be adequately trained. British Standards 6463[3] and 1011[4] give full details.

3.6   REMOVING HAZARDOUS LIQUIDS
Liquids should be pumped to other process units or to storage. Final cleaning out of a vessel will depend on the nature of the material in it, but it may include flushing out with a neutralising agent, a detergent solution, or water — or a sequence of these. In some cases, it may be necessary to fill the vessel with the cleaning liquid, either at ambient or at elevated temperature, and allow it to soak or be boiled out. The final wash will usually be with water.

Careful consideration should be given to the load-bearing capabilities of the vessel's supports and foundations, since the total weight of water in it may exceed that of its normal contents.

Entry of air during the removal of a flammable liquid must be prevented if there is a possibility of pyrophoric residues being present in the vessel. Inert gas should be admitted until the residues can be rendered harmless by soaking them with water.

3.7 THE REMOVAL OF VAPOUR AND GAS

3.7.1 PIPEWORK

Any liquid in the pipework should first be removed. This removal, if it involved flushing out, may well have displaced the vapour or gas. If not, steam or inert gas purging will be required.

Special care is needed where it is necessary to open a flange to isolate a line, where the presence of a toxic gas or vapour is suspected. Positive pressure breathing apparatus and protective clothing must be worn by operatives in such circumstances.

Lines which have contained high flashpoint liquids must be thoroughly steamed out, since any residue left inside the pipe could form a flammable mixture with air when vaporized or thermally decomposed, eg by the heat from a welding or flame-cutting torch.

To prevent sparking on bridging insulating flanges, or contact with earthed equipment, the power supply to a cathodically-protected pipeline must be disconnected for at least 24 hours in order to allow time for depolarization. The pipeline must be bonded to earth before starting work.

Care should be taken to prevent any flammable liquid leaking on to a hot surface or on to unprotected lagging. The auto-ignition temperatures of some substances are surprisingly low[5], and when certain liquids, eg some heat transfer oils, are soaked into lagging, their auto-ignition temperatures can be reduced by as much as 100°C.

3.7.2 FIXED ROOF TANKS

As much of the liquid in the tank as possible should be removed; first, through the normal tank suction line and finally from the clean-out sump or drain point. The tank should then be isolated and ventilated. Under no circumstances should a side manhole be opened at this stage, since an escape of vapour into the bund area could result, until an air- or steam-eductor (bonded to the tank) has been fixed to a roof manhole. After ensuring adequate venting, to avoid collapsing the tank, the eductor can be brought into operation.

Ventilation should be carried out until safe readings (below 5% of the lower flammable limit) are obtained at all test points. The ventilating system must continue in operation whilst persons are inside the tank. Positive-pressure breathing apparatus should be worn by personnel when removing manhole covers.

When carrying out these operations, the whole area inside the tank bund should be considered as a Zone 1 area, and only explosion-protected electrical apparatus which is suitable for this application should be used.

NB Tanks which contain high flashpoint liquids should generally be treated as described above, since they may produce a flammable vapour in conditions of hot weather and, in any case, the provision of proper ventilation is a prudent safety measure.

3.7.3   FLOATING ROOF TANKS

Before emptying and gas-freeing these,their roof supports should be extended to give maximum headroom for persons working inside.

The roof drain valve should be removed or chained fully open. The drain system on the deck should be free of debris and be operating satisfactorily. The tank may then be emptied of liquid and isolated. Vapour-freeing is then carried out as for a fixed-roof tank, but with one important difference: the ventilation flow is in the opposite direction, since the accumulation of a dangerous vapour/air mixture above the roof must be prevented. The eductor is therefore fitted to a manhole low down in the side of the tank, so that fresh air enters through the roof fittings and seal. It should be ascertained that atmospheric conditions are suitable for dispersing the vapour, since it will be emitted at a much lower level than in the case of a fixed roof tank.

It is important to ensure that no liquid has been retained in pontoons or double decks. These should be ventilated and tested individually. The roof supports and its water drainage system should be checked and flushed out with water.

3.7.4   SMALL TANKS

Tanks of less than about 55 cubic metres capacity can usually be steamed out after removing the liquid and flushing with water. Low pressure steam should be used, with the steam connection bonded to the tank. Manholes, vents and drains should be open to prevent overpressurization. Condensate should be drained off continuously and checked until all traces of the tank contents have gone. When steaming out has been completed, precautions should be taken to prevent a vacuum forming.

If the tank contains a flammable liquid, the same precautions as for fixed roof tanks which contain a low flashpoint liquid must be taken. If the tank contents are toxic, they should be steamed out into a closed system and the interconnecting pipe blanked off when this has been done, to prevent recontamination. Overpressurization and vacuum formation should be guarded against.

NB Steaming out larger tanks is likely to be ineffective, since it is difficult to fill the tank with steam , particularly in cold weather.

3.7.5 PROCESS VESSELS

The above methods for cleaning out tanks also apply in many cases to process vessels but they may need to be adapted to suit the particular materials in them, or special process conditions. Steaming out or inert gas purging will often be sufficient. A good general principle is that all flammable vapours and toxic gases should be displaced with steam or inert gas before admitting air. During purging, care should be taken that no pockets of vapour remain in downcomers, behind baffles, or in similar places.

3.8   REMOVAL OF SOLIDS

Two distinct types of operation are involved. The first is where process materials are to be removed from vessels in which they are normally contained (eg bins, silos, reactors and absorbers). The second is where solids such as sludges and residues remain as a contaminant from the process.

If the material to be dealt with is in the form of a finely divided powder, the possibility of a dust explosion could arise. This can be dealt with by steaming out, washing with water after emptying the vessel or purging with inert gas. Care should be taken to avoid the production of dust clouds when using steam or water jets.

3.8.1   PROCESS SOLIDS

These include reactants, catalysts, adsorbents, and final products.

Solid reactants should be fed forward through the process until the vessel is empty. Catalysts, adsorbents and final products should be transferred to suitable containers after liquid and vapour removal has been completed. Special precautions will be needed with toxic, flammable or pyrophoric materials. Radioactive substances are subject to statutory controls.

Activated carbon and other adsorbents should be desorbed, if practicable, before maintenance work starts, although this cannot always be done completely, and some flammable or toxic material can remain. Sometimes, in the presence of air, spontaneous ignition may then occur.

3.8.2 SLUDGES AND DEPOSITS

In a vessel which has contained a flammable or toxic substance, testing may indicate a safe atmosphere after liquid and vapour removal but disturbance of any solid residue may produce a dangerous gas or vapour. Therefore, full protective clothing and breathing apparatus must be used by operatives, and atmospheric testing continued, whilst the solid is being removed. Wooden scrapers and shovels, as well as Zone 1 electrical apparatus, should be used if there is any risk of flammable vapours being present.

Deposits which have been formed in the absence of air may be pyrophoric in which case they may, when dry and exposed to the air, rapidly oxidize, glow, and initiate a fire or explosion. A common pyrophoric substance is one form of iron sulphide; this can be formed by the reaction between iron and hydrogen sulphide, other sulphur compounds, or sulphate-reducing bacteria. Pyrophoric substances are safe when wet, so they should be kept in this state until they can be disposed of safely.

Soot and ash deposits in furnaces, chimneys and flues may contain vanadium pentoxide, which can have toxic effects on a worker after prolonged exposure to it. An approved dust respirator will provide adequate protection if the atmosphere is gas-free and has adequate oxygen (above 20%).

If hot-work is to be carried out on a tank which may contain a heavy residue, special precautions are necessary, see Section 4.3.3.

The removal of liquid-contaminated lagging is considered here as if it were a hazardous residue. As mentioned above, the auto-ignition temperatures of liquids which have soaked into lagging can be much lower than normal. Exposure to the air may cause a fire if the equipment is not cooled down before removal begins.

3.9   SPECIALIZED CLEANING TECHNIQUES

3.9.1   CHEMICAL CLEANING

This a specialized operation and should only be carried out by trained and experienced personnel. The following points are important:

PREPARATION

- The working area should be closed off to unauthorized persons and warning notices displayed.
- Cleaning operatives must know how to stop the operations in an emergency.
- Cleaning chemicals must be transported in closed, clearly-marked containers.
- Operatives should wear protective clothing, face visors and respiratory protection.
- A supply of running water and eye-wash bottles should be readily available.
- Used cleaning solutions should be safely disposed of.
- The equipment to be cleaned out should be leak-tested before work starts and an emergency supply of neutralizing agent should be at hand.
- Equipment should be capable of being completely drained out.
- The presence of dead legs which would prevent effective neutralization must be considered.

CARRYING OUT THE WORK

- Equipment should be vented to prevent pressure build-up and the cleaning agent introduced in a controlled fashion so that any gas produced is released slowly.
- Gas detectors should be used to check for releases of dangerous gases. Toxic gases should be dispersed; if flammable gases are suspected, sources of ignition must be excluded and air must not be used for agitating the cleaning solution.
- Quantites of cleaning agent introduced and recovered should be measured and any difference accounted for.

3.9.2 HIGH PRESSURE WATER JETTING

This operation can be extremely dangerous and capable of inflicting severe injuries. It must therefore only be carried out by trained operatives, properly supervised.

Should a person be struck by a high pressure water jet, urgent medical attention is essential. It is, moreover, vital to inform the medical personnel of the nature of the equipment which caused the injury so that the correct treatment can be given at once.

High pressure water jetting should not be carried out in vessels which may contain a flammable mixture of vapour and air; static electricity may cause an ignition.

Barriers and warning notices must be provided around the working area, and access by persons other than members of the specialist team strictly prohibited.

When cleaning out a tube by this method, a substantial barrier should be placed at the outlet end of the tube to give protection against flying debris. Where water jetting is to be done above ground floor level in a plant, any openings in the floor (eg gratings) should be covered over to prevent the jet penetrating to the floor below. When cleaning underground drains and the like, care should be taken in the vicinity of manhole openings.

Operatives should wear waterproof chemically resistant suits, rubber boots with steel toecaps, a safety helmet, gloves and eye protection.

High pressure water jetting should not be commenced until all appropriate precautions have been taken and a competent person has inspected the site and given permission for the work to proceed.

3.9.3   CLEANING BY GRIT BLASTING

In this process, finely divided abrasive material is propelled by a high velocity jet of air on to the surface to be cleaned. It is the preferred method of preparing steel surfaces for receiving protective coatings, and it is essential when epoxy resins or styrenated resin bonded glass fibre laminates are to be applied.

PRECAUTIONS

• Warning notices should be displayed in the vicinity of the operation.

• Operators should wear suitable protective equipment for their eyes, hands and feet and also respiratory protection.

• When grit blasting in a confined space or enclosure, the operator should wear an air-fed helmet and the exit from the enclosed space be kept clear whilst the work is in progress.

• The build-up of static electricity is common in this work. The air hose (which should be electrically conducting), the compressor and the work-piece should all be earthed.

• Substances which contain sand or free silica should not be used for grit blasting.

• The discharge nozzle of the grit blaster should be fitted with an automatic cut out.

• The air intake to the equipment should be situated in an area which is free from flammable or toxic substances.

3.10   ACCESS

3.10.1   GENERAL
A safe and properly maintained route to and from every place where any person has to work at any time, must be provided. The fact that personnel only have to go to a particular place on rare occasions does not provide an exception to this legal requirement. The means of access may be a temporary one, eg a ladder, so long as it is readily available when required and is adequate for its purpose.

3.10.2   STAIRS
These must be of sound construction and properly maintained. They must be kept free of obstructions and of any substance likely to cause a person to slip. All stairs inside buildings and all exit stairs must have a hand rail and also a lower rail or similar protection.

3.10.3   GUARD RAILS
If a person could fall more than 2 metres from any stair, gangway, or working platform, guard rails must be provided.

3.10.4   SCAFFOLDING
Statutory requirements relating to scaffolding are given in the Construction (Working Places) Regulations 1966 and the Factories Act 1961. Useful advice is found in an HSE leaflet[6].

The three basic types of scaffolding are:

• light scaffolding — for men and tools only, up to a loading limit of 15 lb/sq ft.

• general purpose scaffolding — for men and materials, up to a loading limit of 37 lb/sq ft.

• heavy scaffolding — for men and materials, up to a loading limit of 60 lb/sq ft.

Adequate toe-boards should be provided on all scaffolding and platforms.

A scaffolding contractor *must* be told the purpose for which a scaffold is to be used, and he is legally responsible for complying with statutory requirements when it is erected. Thereafter the company using the scaffold should ensure that it is inspected by themselves, or on their behalf, and that it conforms with the requirements of the law whenever it is in use.

All scaffolds from which persons may fall more than two metres should have been inspected within the preceding seven days by a competent person, unless no part of it has been erected for more than seven days. Likewise, such scaffolding should always be inspected, before it is used, after exposure to severe weather conditions. Results of inspections should be properly recorded.

3.10.5 PORTABLE LADDERS

A ladder should be the correct length for the job. If it is to be used as a means of access or as a working place, then it should rise at least 3 ft 6 in above the place of landing, or above the highest rung to be reached by the feet of the person on it.

Metal ladders should never be used near live electrical conductors. Aluminium ladders should not be used where they may be splashed by caustic alkalies; nor in areas where flammable gases or vapours may be present, because of the risk of a source of ignition being provided by aluminium smears on rusty iron or static electricity from a steam leak.

Ladders should be regularly examined for damage, cracking and general wear and tear.

When in use, the foot of the ladder should be on flat level ground and be secure; wedges, etc should not be used. It should be at a safe angle, with the stile reinforcements on the underside. It should be lashed securely or, if this is not possible, be held firmly at the base (footed).

When on a ladder, a person should have one hand available to hold on (unless a safety belt is provided); when ascending or descending, the ladder should be faced and held with both hands. Tools and materials should be carried in pockets or in a container which is slung over the shoulder, or be raised to the place of work with a handline. A ladder should not be used as a place of work except for easy jobs of short duration.

NB Detailed information on ladders is given in British Standards 1129[7] and 2037[8], and in an HSE leaflet[9].

3.10.6 SAFETY ON ROOFS

Many accidents have occurred when persons have fallen through or from the roofs of buildings. A roof structure is often not strong enough to support the weight of a person and properly designed and constructed crawling boards should be used to provide safe access to a roof area. If frequent access is required to a roof then it is worth fitting permanent walkways with guard rails and fixed ladder access to these.

Permanent warning notices should be fixed on all buildings which have fragile roofs. An HSE publication gives useful advice on the subject[10].

3.10.7   BOSUN'S CHAIRS
These should only be used when the work to be done is of short duration or is in a place where the use of a proper scaffold is impracticable.

They should be installed and used only under competent supervision. See also BS 2830[11], the Construction (Lifting Operations) Regulations 1961 and an HSE leaflet[12].

3.11   ATMOSPHERIC TESTING INSIDE EQUIPMENT

3.11.1   GENERAL

Once any hazardous liquid or vapour has been removed from an item of plant, the effectiveness of the operation must be checked. The atmosphere inside must be safe for persons to enter. It must be ascertained that the atmosphere is not flammable/explosive, toxic, asphyxiating or oxygen-enriched.

The instruments used for carrying out the tests should be suitable for their purpose and regularly inspected, calibrated and properly maintained by a competent person. It remains, however, the responsibility of the user to ensure that the instrument is in proper working order.

It is essential that all tests carried out are fully representative of the atmosphere inside the vessel. Disturbance of deposits or sludge may release trapped gases or vapours rendering frequent retesting essential.

3.11.2   FLAMMABLE ATMOSPHERES

At any given temperature and pressure, there are two limits of flammability of a gas or vapour. The lower flammable limit (LFL) corresponds to the minimum concentration of the substance in air necessary for combustion to occur; it is this parameter which is appropriate to the testing being considered. The upper flammable limit (UFL) is the maximum concentration at which combustion can take place. In practical fire and explosion prevention the UFL is really only of academic interest. Only between these two limits can self-propagation of combustion take place once ignition has occurred.

The methods described above for removing flammable substances from plant aim at reducing the flammable gas or vapour concentration well below the LFL. Combustible gas detectors (explosimeters) are one of the commonest means employed for testing the effectiveness of the removal operation. More information is given in an HSE note[13].

An explosimeter will not, however, indicate in the absence of oxygen, even though a flammable gas is present (eg in a vessel which has been nitrogen-purged). Similarly, if the atmosphere is substantially above the LFL, or even the UFL, the meter will initially go over to full scale and then rapidly drop back to a low reading or zero because any oxygen present is rapidly consumed initially and so further oxidation cannot occur — this can happen so quickly that it can be missed. A recheck should be carried out after flushing out the instrument with clean air; if the same result is obtained, then the atmosphere is probably well above the LFL. Special air dilution tubes are available from the instrument supplier to be used in such situations.

Various types of explosimeter are available. General purpose instruments will detect most flammable vapours and gases, but must be calibrated for the particular one which is being measured. Specific instruments must be used, for example for hydrogen and acetylene. Some substances reduce the sensitivity of the explosimeter or may prevent its working at all, eg lead alkyls, hydrogen sulphide, and high-boiling oils. This problem can be overcome, but needs the assistance of the instrument manufacturer, with whom any unusual applications should be discussed before purchasing it.

If testing shows that the concentration of flammable gas in the atmosphere of a working area is below 5% of the LFL, it then has to be decided if persons can work there without breathing apparatus — this can never be so in a confined space — and can introduce sources of ignition. If sludge is present in a vessel, or if it has a lining, it must never be assumed that the gas concentration will remain at a safe level; frequent atmospheric testing must continue to be carried out.

If the atmospheric concentration is above 5% of the LFL, then work should not normally be carried out, unless special precautions are taken.

3.11.3 TOXIC ATMOSPHERES

The effects of toxic fumes can be considered as short term (safety) and long term (health). In plant maintenance work, it would be expected that the first of these is the one which is mainly relevant.

Short term exposure limits for various substances are given by Mackinson[14]. The unit used is the Immediately Dangerous to Life or Health concentration, which is defined as the concentration from which a person can fully recover after 30 minutes exposure and which does not impede escape, which is particularly important when working in confined spaces.

NB For substances which are both flammable and toxic, eg many industrial solvents, the lower flammable limit is invariably many times higher than the toxic concentration. It must not therefore be assumed that, if the atmosphere has been diluted to below a dangerous flammable level, a toxic risk does not still exist.

Testing for toxic substances can be quite simple or, upon occasion, sophisticated methods will be needed. A convenient system, of wide and easy application, is the chemical detector tube (eg the Draeger tube). In this method an accurately measured volume of air is drawn through a tube which contains a chemical reagent which is specific to the gas being tested, using a calibrated bellows. The reagent becomes stained, and the length of the stain represents the concentration of contaminant in the air.

Other, more complicated, methods of testing are given in a series of booklets published by the HSE[15]. If a specific substance is not covered by these, samples of air may be tested in the laboratory using gas chromatography, infra red spectroscopy, etc. In some situations, on-line analysers which are normally used for process control can be provided with additional sample points so as to monitor specific items of equipment.

3.11.4 OXYGEN TESTING

This is used to guard against (a) oxygen depletion, which can cause death by asphyxiation and (b) oxygen enrichment, which can be a serious fire risk.

Portable oxygen analysers are available for testing for oxygen deficiency or enrichment.

3.12 EQUIPMENT WHICH IS SENT OFF SITE

Owners of plant have a responsibility under the HSW Act to advise third parties (as well as their own employees) who might be involved in the repair or maintenance of their plant. The following points should taken into account:

• Is the item of equipment free from dangerous materials?

• If it is not, adequate warning and instructions must be given about the appropriate safety precautions to be taken.

• Adequate instructions must be given about the identity of any contaminants and their safe disposal, once removed.

Every effort should be made to decontaminate the equipment as thoroughly as possible before it leaves the site. Where this cannot be done as completely as one would wish, the third party must be fully advised about any residual hazard.

4. CARRYING OUT THE MAINTENANCE WORK SAFELY

4.1 INTRODUCTION

Maintenance work on a process plant may be dangerous because of the hazardous substances which are contained within it or which may be in the working environment.

Before work begins on plant or equipment, all persons who will be involved should be fully instructed on its condition with respect to its state of isolation from other parts of the process and also on all potential hazards likely to be encountered. The HSW Act requires that employers provide a safe place and safe systems of work. The development and use of a system of documentation which gives a clear indication of the state of the plant before, during and after work is done on it is a essential aid to safety. Details of suitable systems are given later.

4.2 SUPERVISION

Dangers which might be encountered during maintenance work can be minimized if all those who are involved are properly trained, experienced, competent and, above all, effectively supervised. Those who control the work should be familiar with all the engineering, process and environmental aspects of it and the relationship between these and the personnel and their working methods. They should also be aware of the relevant statutory regulations and technical codes of practice which are to be used to ensure that the work is safely carried out. Inspectors of the HSE may be able to give guidance on this, or a qualified consultant may be employed, if the knowledge is not available in-house.

4.3 WELDING AND FLAME-CUTTING

4.3.1 TOXIC RISKS

Fumes from welding and flame-cutting can be a health risk due to their toxicity and special precautions are necessary when carrying out such work. Particular care is needed when carrying out hot-work on lead, zinc, copper, cadmium, or metals coated with these substances (even as a paint). Welding with low

26

hydrogen alloy or stainless steel electrodes is a particular hazard due to the evolution of volatile fluorides.

Hot-work which is carried out in an ill-ventilated working area may, according to the circumstances, give rise to an oxygen deficient atmosphere and such work should not be carried out unless adequate ventilation (proved by tests on the atmosphere) has been provided. If this is not practicable, then breathing apparatus must be worn.

4.3.2 FIRE AND EXPLOSION RISKS

Connections and repairs to plant which is under pressure should only be done when all alternative methods have been considered. These operations can be carried out if the correct precautions are taken, eg by reducing the pressure, maintaining the flow of fluid or preventing burn-through. The work must be done by skilled operatives under close supervision and with a formal written procedure. The appropriate safety provisions, eg fire-fighting equipment and protection against toxic hazards, should be in place. The fitting of branch connections to equipment under pressure is referred to as 'hot tapping'. The method consists essentially of either bolting or welding a flanged branch to the wall of the equipment and then drilling or trepanning through it into the latter.

Under-pressure repairs include the overplating of thinned sections of equipment and the capping-off of leaks; both these operations involve welding on to the parent equipment.

Hot-work should not be carried out on any plant which contains:
* any mixture of gases or vapours within the flammable range or which may become flammable as the result of the welding work;
* any substance which may react or decompose giving a dangerous increase in pressure or an explosion which may cause failure of the metal, see Section 4.3.3;
* hydrogen, or mixtures containing hydrogen, where the partial pressure of the hydrogen exceeds 7 bar;
* oxygen or an oxygen-enriched atmosphere which may cause 'runaway' burning of the metal;
* compressed air together with hydrocarbons or other combustible materials.

4.3.3 HOT-WORK ON TANKS CONTAINING HEAVY RESIDUES

Many explosions have occurred in tanks which contain heavy, non-volatile residues, oils and polymers, which do not produce flammable vapours at ambient temperature and so would not show a positive test with an explosimeter. It is

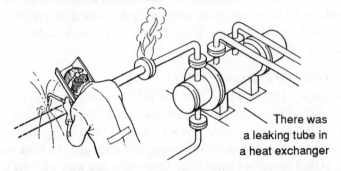

An empty water line caught fire while a welder was working on it!

There was a leaking tube in a heat exchanger

Test the inside of water, air, steam and nitrogen lines before welding on them in a plant area

almost impossible completely to clean out a tank which has contained such materials. This is particularly so if the tank is corroded so that material can get between the plates where there is a defect in the welding. Some older tanks are, in fact, welded along the outside edge of the lap only. Residues can also accumulate behind tank linings.

When heat is applied by a welding or cutting torch to the outside of a tank, the residue is thermally decomposed to produce flammable gases. These mix with the air in the tank to form a flammable atmosphere which can be ignited and cause a serious explosion.

It may be difficult to remove solid residues from a vessel by steaming it out. Where the vessel or pipework is too small to be entered for cleaning, hot work may be carried out after filling with water, inert gas or inert gas-filled foam. This technique of filling vessels, etc with nitrogen-expanded foam gives a very high standard of safety and should be employed in difficult situations where thorough cleaning out and the provision of a safe atmosphere inside is difficult to achieve.

4.3.4   GENERAL PRECAUTIONS

In no circumstances should work proceed if equipment will be operated outside its design rating for temperature and pressure. The HSE publishes a valuable guidance note on carrying out hot-work safely on process plant[16]. A plan for personnel evacuation must be prepared before work commences and provision made for it to be applied in any emergency which may arise whilst carrying out the work. In all situations, easy means of access and escape must be provided and congested working areas avoided.

To avoid stray currents when arc welding, the welding return lead should, where possible, be directly attached to the equipment being welded. Welding cables should not come into contact with pipework and, where they must cross it, they should do so over a suitable bridge or go underneath and be protected against contact.

The normal precautions with hot-work and the requirements of any Certificate of Exemption must always be observed. In particular:

• Any substance inside the equipment should not cause metallurgical changes in the heated area.

• The in-service temperature of the equipment should be considered.

• Post-welding heat treatment requirements and any special treatments or welding rods must be specified.

• Consideration should be given to the thickness of the metal in the area to be welded, and its ability to withstand stress conditions and to receive any connection and support material. Welding should not normally be carried out where the metal is less than 6 mm thick, because of the risk of burning through it.

• Where it is necessary for safety, an adequate flow of process material through the plant must be reliably maintained.

4.4   DANGER IN OXYGEN-ENRICHED ATMOSPHERES

Oxygen concentrations in the atmosphere above the normal 21% can be very dangerous, particularly in confined spaces. Materials such as fabrics used for clothing, which might only smoulder when ignited in a normal atmosphere, will burn vigorously when the oxygen content has been increased by only a few percent. Thus stringent precautions must be taken to prevent this happening.

Gas-welding and cutting equipment should be removed from a confined space when not in use; when it is being used, a high standard of positive-pressure general ventilation of the space should be provided since, particularly when flame-cutting, a large excess of oxygen may be liberated into the atmosphere causing dangerous enrichment. Compressed gas cylinders should never be taken into a confined space.

4.5   WORKING IN CONFINED SPACES

Serious accidents frequently occur when persons are working in confined spaces. Some of these places are quite obvious, such as reaction vessels, tanks, large ducts and culverts, sewers and drains. Other, less obvious, examples include open-topped tanks, vats, furnaces and ovens.

A toxic or flammable atmosphere can arise in a confined space either from a gas or vapour left behind from a process; one which enters the space before or during the maintenance work due to incomplete isolation; fumes from the disturbance of a sludge or a tank lining; or from a process being carried out inside, eg painting or welding. Oxygen depletion or enrichment of the atmosphere in the space may also be a serious danger.

To work safely in a confined space requires a strict system of precautions under the control of a permit-to-work system. Preparation for vessel entry has been dealt with above and, when this has been properly done and certified, the specified precautions to be taken whilst working in the confined space must be adhered to.

These precautions are likely to include:

• general ventilation of the space, usually by supplying positive pressure air with, sometimes, additional local exhaust extraction;

• the provision of positive pressure breathing apparatus;

• the availability of equipment and trained personnel for rescue purposes whenever someone is inside the confined space;

• efficient maintenance of all the above equipment;

• ensuring safe and easy access into and exit from the confined space, particularly with respect to the size of manhole and other openings;

• efficient communication with persons inside the confined space is essential; a telephone or two-way radio may be appropriate.

In addition to the normal precautions for entry into a confined space, additional ones may be required when entering sewers, drains and other water systems. The skin should be protected by means of suitable clothing, rubber boots and gloves. Respiratory protective equipment, life-lines and rescue equipment may be required. Open wounds should be covered with a sealed dressing.

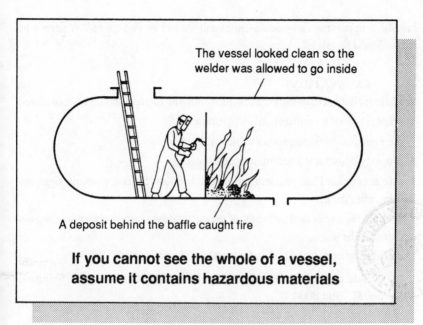

The vessel looked clean so the welder was allowed to go inside

A deposit behind the baffle caught fire

If you cannot see the whole of a vessel, assume it contains hazardous materials

All personnel and equipment should undergo thorough cleaning when the work has been completed.

When a blockage is to be removed from a system, the possibility of a sudden rush of liquid from behind the blockage should not be overlooked. All electrical apparatus and communications equipment should be waterproof and be suitable for use in an explosive atmosphere, unless it can be positively established that the system cannot contain flammable gases or vapours.

Guidance on safe working in confined spaces is given in an HSE note[17].

4.6   WORK IN BUNDED AREAS

Special precautions are necessary whenever work is to be done in a bunded area. Dangerous concentrations of flammable or toxic fumes may accumulate in a poorly ventilated open area which is surrounded by a wall which is more than about a metre high.

Entry of persons into a bunded area should be generally thought of as being similar to entry into a confined space, eg a process vessel, and the need for precautions such as atmospheric testing, ventilation and the provision of

31

breathing apparatus and rescue equipment should be considered. A permit-to-work should be issued.

4.7   EXCAVATIONS

When carrying out excavation work, the following matters should be considered:
* preparation of a certificate of excavation;
* the presence of underground cables and pipework;
* properly constructed and maintained shoring;
* safe access, and the provision of secure barriers, fences, warning signs and lighting after dark;
* keeping the excavated soil clear of the excavation to prevent overloading and collapse of the sides;
* monitoring and control of water seepage.

Safe practices in excavation work are described in British Standards 5930[18], 6031[19] and 8004[20].

4.8   CARE WITH GAS CYLINDERS

Cylinders of compressed gases are often encountered in plant maintenance work and dangers may arise from their use if care is not taken. The more common gases are acetylene, propane and hydrogen (flammable and explosive) and oxygen, nitrogen and argon (non-flammable but still potentially dangerous).

Where practicable, cylinders should be stored in a properly constructed secure compound out-of-doors. Full and empty cylinders should be segregated and the different types clearly identified. Data sheets which give guidelines on the safe storage of compressed gas cylinders are available[21]; for LPG (eg propane) cylinders, there is advice in an HSE publication[22].

When in use, compressed gases are obviously more potentially dangerous than when they are in safe storage. Guidelines for the safe use of compressed gases are given in an IChemE publication[23].

Gas cylinders other than those used for breathing apparatus should never be taken into a confined space and oxygen should never be used to 'sweeten' the atmosphere of such a place.

Cylinders, torches and hoses must be removed from the working area at the end of the operation or shift, or if the work is temporarily suspended. Empty gas cylinders must be removed at once.

4.9   TOXIC SUBSTANCES

These may enter the body by inhalation, skin absorption or ingestion and their effects may either be immediate (eg gassing by carbon monoxide) or long term (eg lead poisoning). Examples of situations where problems may occur are given throughout this booklet, together with the appropriate precautions to be taken. Information on both the short and long term toxic effects of a wide range of substances is given an HSE publication[24].

An important piece of legislation which deals with the protection of workers against risk to health is the Control of Substances Hazardous to Health Regulations 1988. These require (among other matters) that an assessment of this risk be carried out as part of the pre-planning of all maintenance activities. Control measures should be assessed, including the provision of personal protection. Employees should be consulted about how the work is to be done and about any potential problems which may be encountered. The assessment should reach one of the following conclusions:

• Risks to health are insignificant, so no further action is needed except to record the assessment and review it as circumstances change.

• Risks to health are adequately controlled provided that certain control measures are maintained.

• Risks to health are significant and are not adequately controlled. Interim measures for preventing or controlling exposure must be identified and further action specified.

4.10   ASPHYXIATION

Asphyxia is the condition in which the supply of oxygen to the lungs is reduced below the body's requirements. Nitrogen, carbon dioxide and many hydrocarbon gases are simple asphyxiants and are distinguished from the toxic gases, eg carbon monoxide.

The risk of asphyxiation is not uncommon when carrying out maintenance work on process plant, as shown by the following examples. When an empty mild steel vessel has been closed up for some considerable time, the oxygen in it may be consumed by reaction with the steel to form rust; welding work in a confined space may reduce the oxygen content of the atmosphere; inert gas purging of a vessel will obviously remove oxygen; the admission of a fuel gas such as methane into an oven or furnace is dangerous if persons are inside at the time, and active carbon absorbers may be deficient in oxygen.

Air normally contains 21% oxygen. Distress will be experienced when this falls to around 16% and collapse and probable death below 10%. Entry into a confined space should not be permitted without breathing apparatus if there is any foreseeable risk of this happening.

Adequate positive-pressure general ventilation of a working area — particularly if it is a confined space — will be required to ensure that the oxygen level is safe. If oxygen deficiency is likely to be an ongoing problem, eg if hot-work is being carried out, then atmospheric testing should be carried out throughout the duration of the operation.

4.11 ASBESTOS

A material which is particularly hazardous to health, which can be encountered in plant maintenance work and which merits special attention, is asbestos. In the context of this book, the most likely operation to be met with will be the stripping of old asbestos insulation from pipework and process equipment.

Because of the serious health risk, stringent precautions are required by law whenever exposure of personnel to asbestos is possible, and so asbestos stripping is usually carried out by specialist contractors who are required to possess a licence, which should ensure that they are properly equipped for the task.

The following legal regulations are relevant:

- Control of Asbestos at Work Regulations, 1987
- Asbestos (Licensing) Regulations, 1983
- Asbestos (Prohibitions) Regulations, 1985

These specify the legal requirements when working with asbestos. Full and detailed guidance is given in a number of HSE publications[25, 26, 27, 28].

4.12 IONIZING RADIATIONS

The control of the exposure of workers to ionizing radiations is a very specialized subject and expert advice should always be obtained before carrying out work where a potentially hazardous situation may arise. The legal requirements are given in the Ionizing Radiations Regulations, 1985.

4.13 EXPLOSIVES

It is not felt to be within the scope of this booklet to give detailed advice on the safe use of explosives in an industrial situation, eg for blasting or demolition. Assistance may be obtained from HM Inspectorate of Explosives which is part of the Health and Safety Executive, through whom it may be contacted.

Cartridge operated tools are commonly used items of industrial equipment and, depending as they do on an explosive charge, require much care in using them safely. It is a statutory requirement that they shall only be used by or under the immediate control of a competent person who has adequate knowledge of the dangers associated with them. Detailed advice on the matter is found in an HSE guidance note[29].

4.14   PRESSURE TESTING OF PROCESS VESSELS AND EQUIPMENT

4.14.1   NEW EQUIPMENT

Newly fabricated vessels, pipework, etc should always be tested after reference has been made to the appropriate code of practice and design data upon which it has been constructed, to obtain details of the test pressure.

4.14.2   MODIFIED OR REPAIRED EQUIPMENT

If a repair or modification, eg welding, is carried out on an item of plant which may significantly affect its strength, it should be pressure-tested before being returned to service. Where pressure testing is not reasonably practicable, more stringent non-destructive testing should be employed than would otherwise be the case.

4.14.3   TESTING METHODS

Testing should be hydrostatic and, where water must be excluded, other fluids may be used. Care must be taken with the quality of the water used in testing. For example, nitrates and chlorides in it may affect stainless steel. Precautions are needed if there is a risk of freezing during the test.

Pressure gauges must be calibrated before and after use, and the effect of the static head must be taken into consideration.

Care should be taken during the filling of the vessel to avoid overpressurization by mains pressure or by thermal expansion. Expansion bellows should be restrained and be able to withstand axial thrust. When emptying, the equipment must be properly vented to prevent a vacuum forming.

Pressure testing should be carried out in accordance with a laid down written procedure.

4.14.4   PNEUMATIC TESTING

In some cases, due to design or operational restraints, hydrostatic testing may not be possible. In such cases, pneumatic testing may have to be considered.

Unless this is done properly, it can be very dangerous. It should be carried out only with the involvement and agreement of the Inspector of the HSE.

4.14.5   LEAK TESTING

When an item of equipment, eg a process vessel, has only been opened at a bolted or screwed connection, and then reassembled, the joint should be leak tested after remaking it. The test should be carried out at a pressure below the design pressure of the vessel, as its primary purpose is simply to test the remade joint for leakage.

5. PERMITS-TO-WORK AND RELATED DOCUMENTATION

5.1 INTRODUCTION

The Factories Act 1961 requires employers to provide a safe means of access to a place of work which is itself safe and, under certain circumstances, authorization and certification of the work. But, much more relevant to this chapter, the HSW Act requires that systems of work should be provided and maintained which are safe and without risk to health.

Against this background, the chemical and allied industries have developed procedures and supporting paperwork designed to meet their own particular safety needs; a practice which is gradually being taken up more widely, in other industries. It is essential that all maintenance work, however minor, is properly authorized, controlled and recorded; a permit-to-work system is an acceptable means of meeting this requirement. If, however, the system is too complex, or is used unnecessarily, it can fall into disrepute and its value is thereby diminished. To avoid this, repetitive work which can be precisely defined is better carried out under the control of its own system of documentation, rather than being drawn into the general permit-to-work system on every occasion upon which the work is done.

5.2 CONTENTS OF A PERMIT-TO-WORK

Apart from legal aspects, for a permit-to-work to be able to meet real operational needs, it must include certain essential information, as follows:

• The name and address of the Company to which it belongs and, if it is different, a reference to the premises where it will be used.

• An individual and exclusive reference number.

• The date upon which the work may be done and the starting and finishing time. The 24 hour clock should be used to avoid confusion.

• Positive and unambiguous identification of the location of the job and the equipment to be worked on. The use of colloquial terms should be avoided, or included only in a supplementary role. As an example:

Correct location description: 'Building No.5'

There were seven pumps in a row

A fitter was given a permit to do a job on No.7
He assumed No.7 was the end one and dismantled it
Hot oil came out
The pumps were actually numbered:

1   2   3   4   7   5   6

Equipment which is given to maintenance must be labelled
If there is no permanent label then a numbered tag must be tied on

Colloquial name: 'Redundant plant store'
Correct equipment description: 'Serial No.8081E'
Colloquial name: 'Battery charger'

• Clear details of the work to be done, by what means and the number of persons covered by the permit.

• The safety precautions to be provided before and during the operation, shown in such a way as to reduce the risk of alteration or further insertions. General statements, eg 'full protective clothing' or 'trade competence accepted', should not be used.

The layout of the form should cater for a cessation of work for reasons other than its completion and for the restoration of normal activities, eg failure to complete it within the time allowed or the discovery that the job is going to take longer than planned.

It is usual for forms to be prepared in sets of 2 or 3 copies, a number of sets being bound together into a book. Each copy can be overprinted to give its location after it has been completed, eg 'Display at job location', 'Display in control room' or 'File copy — retain in book'. It is sensible to limit the number of forms available in a department and to use them in sequential order. A record showing the department to which they were issued and the dates of issue and return is valuable. All this implies a single issuing centre.

No specific format for a permit-to-work is given, since its composition will vary from company to company. Some essential and desirable features of a permit are given in Tables 5.1 and 5.2 overleaf.

A common practice in the chemical industry is to pay special emphasis to particularly hazardous types of work, viz. that which requires a person to enter a confined space, or involves the use of naked flames ('hot-work'). This is done by using forms, often of different colours, dedicated to these activities. The issue of such forms is still governed by the principles in this chapter, and they are complementary to each other and to the general permit-to-work system. For example, a welding operation inside a vessel would require a general permit to cover vessel cleaning and isolation, as well as a hot-work permit. Due to the special risks involved in such operations, the number of persons authorized to sign entry and hot-work permits should be less than for the general ones.

These high-risk operations warrant ratification of both the work content and the safety precautions by a member of senior management, and this ratification may include a meeting of all parties involved before work starts. Such

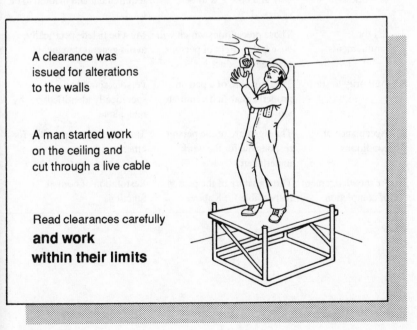

A clearance was issued for alterations to the walls

A man started work on the ceiling and cut through a live cable

Read clearances carefully **and work within their limits**

TABLE 5.1
Essential features of a permit-to-work proforma

Essential requirements	Explanation of requirements	Reason/comment
1 Owner's identity	Company's name and address in full	The form fulfills a legal requirement and may be required as evidence
2 Explanatory title	Name of form	For the purpose of 'in house' identification
3 Proforma identity	Form serial number/book number, etc: 1, 2, 3, 4, 5, etc	There must be positive identification without repetition
4 Allowed time	Date of work to be done and time from and up to	This is a simple control requirement
5 Location of work	Building, plant area, vessel or equipment identification	It must be positive; avoid colloquial references
6 Description of work	What is required to be done, why and how it is to be achieved	Simple statement of requirement and method to be used
7 Safety requirements	Those precautions which will ensure the safety of persons, plant and product	Must be listed specifically; avoid general statements
8 Authority to start	The signature of a person properly qualified to initiate work	Persons should be specifically identified; not by rank alone
9 Acceptance of conditions	The signature of the person responsible for the work undertaken	They are then responsible for others
10 Acknowledgement of completion	The signature of the persons responsible for 8 above	Restoration of normal functions

TABLE 5.2

Desirable features of a permit-to-work proforma

Desirable features of proforma	Explanation of requirements	Reason/comment
1 Cautionary statement	This is a positive reminder to the initiator of the importance to be attached to the document	It is a legally binding document
2 Advisory statement	This reminds the initiator that the responsibility for introducing effective safety requirements is his/hers	The initiator may delegate this work to others but cannot then delegate responsibility
3 Recognition of hazard level	This is a positive method of making the initiator aware of high risk situations	Relates to both place of work and nature of work
4 Acknowledgement of high risk hazard	High risk situations often require and warrant more than a single viewpoint	Consideration by and input from other disciplines warranted
5 Indication of work progress	Commencement of work may often reveal a far greater job content than was originally envisaged	Provides an opportunity for job close down and reappraisal

parties might include process operators, maintenance staff and laboratory technicians. If this procedure is carried out, it must be recorded on the permit.

The authority to proceed with the work is vested in the initiator of the permit. The acceptance of the work, with its regulating conditions, is the responsibility of the recipient. Both the initiator and the recipient should sign the form. Similarly, completion or termination of the work must be formally acknowledged by the initiator, who should then sign again.

Properly prepared permit forms will prompt the adoption of safe working practices but they cannot replace careful consideration of the risk potential on every individual occasion. The use of check lists, etc as part of the printed form will be a matter of individual company choice.

In preparing a permit-to-work, the effects of the work on adjoining parts of the factory, or even places off-site, should always be considered.

**Some works display
'PERMITS-TO-WORK'
on the job**

Everyone can read them

The content of a permit-to-work form should remind the signatories of their responsibilities and of the importance of the document to which they are a party. The recipient should have a full understanding of the nature of the work to be done and the precautions to be taken, when he accepts the form.

Over elaboration should be avoided in preparing a permit-to-work system; it should be capable of being clearly understood, unambiguous and written in a practical way. Unnecessarily bureaucratic procedures should be avoided.

The following pages contain examples of permit-to-work forms of the type used by a number of commercial companies. The purpose of the examples is to demonstrate ways in which the forms may be prepared.

XY CHEMICALS   COLD WORK PERMIT

NOT to be used for entry into a confined space and/or work involving a source of ignition and/or radiation

Serial

Contractor . . . .
Pass no. . . . .
Site contact . . . . . Tel. . . .
Permit office . . . .

1 LOCATION	2 PERMIT VALID FROM TIME
. . . .	OF CERTIFICATION UNTIL
Plant reference Nos.	Time . . . . Date . .
3 WORK TO BE CARRIED OUT	WORK ON ELECTRICAL
. . . .	APPARATUS . . . .
. . . .	. . . .

4a ISOLATION — Details of isolations carried out	Time	Date	Initial
. . . .	. . . . .	. . . . .	. . . . .
. . . .	. . . . .	. . . . .	. . . . .
. . . .	. . . . .	. . . . .	. . . . .
. . . .	. . . . .	. . . . .	. . . . .

As per Plant Maintenance Preparation Procedure No.

4b ELECTRICAL ISOLATION — Please isolate the following: Signed . . . .

Authorised person

Item PRN	Item	Board No.	Circuit No.	Bag No.	Time	Date	Isolated by
. . . . .	. . . .	. . . . .	. . . . . .	. . . . .	. . . . .	. . . . .	. . . . .
. . . . .	. . . .	. . . . .	. . . . . .	. . . . .	. . . . .	. . . . .	. . . . .
. . . . .	. . . .	. . . . .	. . . . . .	. . . . .	. . . . .	. . . . .	. . . . .

5a HAZARDS REMAINING . . . .
. . . .
. . . .
. . . .

5b PRECAUTIONS TO BE TAKEN . . . .
. . . .
. . . .
. . . .
. . . .

6 CERTIFICATION

I certify that the precautions in Section 4 have been taken and that work may start subject to the requirements of Section 5

Signed . . . . Position . . . .
Time . . . . Date . . . .

7 ACCEPTANCE

I have read and understood the permit and accept the precautions to be taken

Signed   Position

1 . . . .
2 . . . .
3 . . . .
4 . . . .
5 . . . .
6 . . . .

8a *WORK COMPLETED — I am satisfied that all work is complete and electrical apparatus has been reconnected.

*PERMIT WITHDRAWN — I have agreed the work still outstanding with . . . .
Delete as appropriate

Signed . . . . Time . . . . Date . . . .

8b ELECTRICAL DE-ISOLATION — Please de-isolate the equipment detailed in 4b

Signed . . . . Time . . . . Date . . . .
Authorised person

EQUIPMENT DE-ISOLATED

Signed . . . . Time . . . . Date . . . .

NB Both Authorised person and electrician should check that it is safe to de-isolate before signing.

9 ACCEPTED AND PERMIT CANCELLED — I have surveyed the work and removed isolations

Signed . . . . Time . . . . Date . . . .

10 COUNTER SIGNATURES

HIGH TEMPERATURES and/or PRESSURE	UNDERGROUND HAZARDS		AWARENESS OF WORKS IN ADJACENT AREAS
	Drawing no . . . . attached shows approx location of underground hazards		
. . . .	. . . .	. . . .	. . . .
Authorised engineer	Nominated draughtsman	Authorised electrical engineer	. . . . Authorised person

NB Authorised person will mark limit of work area on above drawing.

REMOVAL OF ISOLATIONS FOR TESTING BEFORE COMPLETION OF WORK

A. SANCTION TO TEST

NB Before removing isolation for testing the Authorised person must ensure that all personnel working on this permit have been informed that the test is to take place and guards HAVE/HAVE NOT beenreplaced.

Time/ date	Item PRN	Item	Permission to de-isolate	Precautions	Acceptance	De-isolated by
1 . .	. . . .	. . . .	Authorised person	. . . .	7(1) signatory	. . . .
2 . .	. . . .	. . . .	Authorised person	. . . .	7(1) signatory	. . . .

B. ISOLATION AFTER TEST

Time/ date	Authorised person	Isolated by	I certify that the original permit conditions are now restored
1 . .	. . . .	. . . .	Authorised person
2 . .	. . . .	. . . .	Authorised person

RENEWALS — The work area has been surveyed and conditions have not changed. Validity extended until:

Time	Date	Signature	Time	Date	Signature
. . .	. . .	. . . .	. . .	. . .	. . . .
. . .	. . .	. . . .	. . .	. . .	. . . .
. . .	. . .	. . . .	. . .	. . .	. . . .

XY CHEMICALS ENTRY PERMIT   Serial . . . . .

DATE AND TIME OF	From	To	Work site
INITIAL 8 HOUR PERIOD	. . . .		
EQUIPMENT NAME/NUMBER			
DESCRIPTION OF WORK			
. . . .			

MARK: Required ✗ Checked ✓	Yes	No	Ch'kd	MARK: Required ✗ Checked ✓	Yes	No	Ch'kd
A. Has equipment to be:	. .	.	. . .	3 Is fire protection sited?	. .	.	. . .
1. Depressured	. .	.	. . .	4 Is suitable access and exit			
2 Drained	. .	.	. . .	provided?	. .	.	. . .
3 Isolated by	. .	.	. . .	5 Are standby personnel			
blanking? blank list No.	. .	.	. . .	required?	. .	.	. . .
disconnecting?	. .	.	. . .	6 Are lifelines, harness and			
valving?	. .	.	. . .	breathing apparatus required?	. .	.	. . .
4 Steamed?	. .	.	. . .	7 Is air compressor sited			
5 Waterflushed?	. .	.	. . .	correctly?	. .	.	. . .
6 Purged with inert gas?	. .	.	. . .	8 Neutralising solution on	. .	.	. . .
7 Ventilated by natural/				9 HF F/Aid kit available	. .	.	. . .
mechanical means?	. .	.	. . .				
8 Neutralised?	. .	.	. . .	C. Excavations:	. .	.	. . .
				1 Are sides safely stored?	. .	.	. . .
B. 1 Are sewers, drains, etc				2 Is a barrier erected?	. .	.	. . .
within 50 feet of worksite							
sealed?	. .	.	. . .	D. 1 Have repeat gas/toxic/			
2 Is site clear of combustible				oxygen tests to be made?	. .	.	. . .
materials?	. .	.	. . .				

AUTHORITY GRANTED TO ISOLATE . . . . Time . . . . Date . . . .
Issuing authority

EQUIPMENT ISOLATED . . . . Time . . . . Date . . . .
Electrical section

PROTECTIVE CLOTHING TO BE WORN: indicate by 'X'
(OPERATING AUTHORITY)   ALKYLATION UNIT

☐ Full protective	☐ Face shield	☐ Goggles	☐ TEL clothing	☐ Acid area clothing
clothing			☐ PVC	☐ Air fed hood
☐ Dust mask	☐ Air-line BA	☐ Self contained BA	☐ Ear protection	☐ Air fed suit

SPECIAL INSTRUCTIONS . . . .
. . . .

GAS TEST RESULTS		*Combustible/toxic/oxygen			*Combustible/toxic/oxygen		
		Date/time	Result	Signature	Date/time	Result	Signature
	Initial check	. . . . .	. . . .	. . . . .	. . . . .	. . . .	. . . . .
	Re-check	. . . . .	. . . .	. . . . .	. . . . .	. . . .	. . . . .
	Re-check	. . . . .	. . . .	. . . . .	. . . . .	. . . .	. . . . .
	Re-check	. . . . .	. . . .	. . . . .	. . . . .	. . . .	. . . . .
	Re-check	. . . . .	. . . .	. . . . .	. . . . .	. . . .	. . . . .
	Re-check	. . . . .	. . . .	. . . . .	. . . . .	. . . .	. . . . .
	Recheck	. . . . .	. . . .	. . . . .	. . . . .	. . . .	. . . . .

*Delete when not applicable

SITE PREPARATION IS COMPLETE — Permission is granted for work to commence.						
To be signed by issuing authority	Initial check	Date/time	Re-check	Date/time	Recheck	Date/time
	. . . .	. . . .	. . . .	. . . .	. . . .	. . . .
	. . . .	. . . .	. . . .	. . . .	. . . .	. . . .
	. . . .	. . . .	. . . .	. . . .	. . . .	. . . .
	. . . .	. . . .	. . . .	. . . .	. . . .	. . . .
	. . . .	. . . .	. . . .	. . . .	. . . .	. . . .
	. . . .	. . . .	. . . .	. . . .	. . . .	. . . .
	. . . .	. . . .	. . . .	. . . .	. . . .	. . . .
	. . . .	. . . .	. . . .	. . . .	. . . .	. . . .
	. . . .	. . . .	. . . .	. . . .	. . . .	. . . .
	. . . .	. . . .	. . . .	. . . .	. . . .	. . . .
	. . . .	. . . .	. . . .	. . . .	. . . .	. . . .

I am aware that this work is in progress and that conditions are as above.						
To be signed by operator in charge	Initial check	Date/time	Re-check	Date/time	Recheck	Date/time
	. . . .	. . . .	. . . .	. . . .	. . . .	. . . .
	. . . .	. . . .	. . . .	. . . .	. . . .	. . . .
	. . . .	. . . .	. . . .	. . . .	. . . .	. . . .
	. . . .	. . . .	. . . .	. . . .	. . . .	. . . .
	. . . .	. . . .	. . . .	. . . .	. . . .	. . . .
	. . . .	. . . .	. . . .	. . . .	. . . .	. . . .
	. . . .	. . . .	. . . .	. . . .	. . . .	. . . .
	. . . .	. . . .	. . . .	. . . .	. . . .	. . . .
	. . . .	. . . .	. . . .	. . . .	. . . .	. . . .
	. . . .	. . . .	. . . .	. . . .	. . . .	. . . .

PERFORMING AUTHORITY INSTRUCTIONS: . . . .
. . . .
. . . .

I understand the precautions to be taken and have instructed the person/persons carrying out the work accordingly.						
To be signed by operator in charge	Initial check	Date/time	Re-check	Date/time	Recheck	Date/time
	. . . .	. . . .	. . . .	. . . .	. . . .	. . . .
	. . . .	. . . .	. . . .	. . . .	. . . .	. . . .
	. . . .	. . . .	. . . .	. . . .	. . . .	. . . .
	. . . .	. . . .	. . . .	. . . .	. . . .	. . . .
	. . . .	. . . .	. . . .	. . . .	. . . .	. . . .
	. . . .	. . . .	. . . .	. . . .	. . . .	. . . .

	Time	Date	Signed
Work completed	. . . .	. . . .	. . . .
			Performing authority
Electrical supply returned to equipment	. . . .	. . . .	. . . .
			Electrical section

XY CHEMICALS HOT-WORK, ENTRY, RADIATION PERMIT Serial

To be used for entry into a confined space and/or work involving a source of ignition and/or radiation. Refer to SSOs 2 and 17.	Contractor . . . . Pass no. . . . . Site contractor . . . . Telephone No. . . . .

1 LOCATION . . . . Area classification	**2 PERMIT VALID FROM TIME** **OF CERTIFICATION UNTIL** Time . . . . Date . . . . .

3 WORK TO BE CARRIED OUT . . . .

. . . .

4a ISOLATION — Details of isolations carried out	Time	Date	Initial
. . . .	. . . . . .	. . . . .	. . . . .
As per Plant Maintenance Preparation Procedure No.	. . . . . .	. . . . .	. . . . .
Other isolations/preparations carried out on cold work permit No	. . . . . .	. . . . .	. . . . .
4b ELECTRICAL ISOLATION — Details of isolation carried out . . .	. . . . . .	. . . . .	. . . . .
. . . .	. . . . . .	. . . . .	. . . . .

5 ATMOSPHERIC TEST CERTIFICATE — I certify that the working area CAN/CANNOT be entered without breathing apparatus and IS/IS NOT safe for means of ignition as specified in section 6 below

NO GAS TEST DONE/GAS CERTIFICATE ATTACHED

Certificate No. . . . . Signed . . . . Time . . . . Date . . . . .

6 NAKED FLAME CERTIFICATE Type permitted	Fire protection required
Naked flames or other ignition	. . . .
sources ARE/ARE NOT permitted	. . . .

7a HAZARDS REMAINING . . . .

. . . . Radiation permit No. . . . . .

7b PRECAUTIONS TO BE TAKEN (Continue overleaf if necessary) . . . .

. . . .

. . . .

. . . .

. . . .

. . . .

8 CERTIFICATION — I certify that the precautions in sections 4 and 5 have been taken and that work may start subject to the requirements of sections 6 and 7.

Signed . . . . Position . . . . Time . . . . Date . . . . .

9 ACCEPTANCE (Continue overleaf if necessary) — I have read and understood the permit and accept the precautions to be taken.

Signed   Position

1 . . . .

2 . . . .

3 . . . .

4 . . . .

5 . . . .

6 . . . .

10 WORK COMPLETED/PERMIT WITHDRAWN

Signed . . . . Time . . . . Date . . . . .

11 ACCEPTED AND PERMIT CANCELLED

Signed . . . . Time . . . . Date . . . . .

12 COUNTERSIGNATURES AWARENESS OF WORK IN ADJACENT AREAS

Endorsement	Works manager or nominee	Appointed member of safety section	
Entry into confined space	. . . .	. . . .	1 . . . . 2 . . . .
Source of ignition	. . . .	. . . .	Authorised Authorised
Renewal permitted	. . . .	. . . .	signatory signatory

RENEWALS

The work has been surveyed and conditions have not changed. Validity extended until:

Time	Date	Signature	Time	Date	Signature
. . .	. . . .	. . . .	. . .	. . . .	. . . .
. . .	. . . .	. . . .	. . .	. . . .	. . . .
. . .	. . . .	. . . .	. . .	. . . .	. . . .
. . .	. . . .	. . . .	. . .	. . . .	. . . .
. . .	. . . .	. . . .	. . .	. . . .	. . . .
. . .	. . . .	. . . .	. . .	. . . .	. . . .

GAS TEST RENEWAL

The locations to be gas tested are

. . . . for . . . .

. . . .

. . . .

. . . .

NB Exact positions to be clearly defined to eliminate any doubt

The following persons are authorised to witness repeat gas tests associated with this permit:

1 . . . .

2 . . . .

3 . . . .

4 . . . .

5 . . . .

6. MONITORING SAFE WORKING PRACTICES

6.1 CHECKING PERMIT-TO-WORK SYSTEMS

Permit-to-work systems need both unscheduled periodic checks and also continuous monitoring if they are not to become sloppy and ineffective. One method of monitoring is to arrange for permits to be vetted daily.

If this is done properly, with the support of management, the scheme should operate effectively. It is also worthwhile, to avoid inbred bad habits, for someone from outside the department to carry out an audit on the system from time to time.

6.2 STANDARD PROCEDURES FOR PLANT PREPARATION

Permits-to-work cover many types of job. The safe completion of a job depends on the process personnel ensuring that the plant is properly prepared for the maintenance staff, and that no vital safety matter has been overlooked.

The preparation of a plant for maintenance work should be treated like any other process operation, and detailed in written procedures. Checklists itemizing each step should be provided; these prevent an important step being forgotten, and they are of particular value when the preparation work extends over a shift changeover.

This approach to plant preparation enables the hazards involved in a job to be properly assessed. It avoids the situation where just one person — the one issuing the permit — has to accept all the responsibility for identifying the hazards and for specifying the appropriate safety measures. It can also considerably reduce the office work load on the permit issuing staff in cases where a standard operating procedure can be employed.

6.3 UP-TO-DATE PLANT DRAWINGS

On all maintenance jobs, it is essential when isolating sections of plant, that up-to-date pipework and instrumentation (P&I) diagrams are available for the plant before the work commences. It is, moreover, essential that these diagrams be kept up-to-date (see Section 6.6).

6.4 MARKING ISOLATION POINTS

A formal system of numbered tags should be used to identify places such as flanges where blanking plates or other devices have been inserted for the purpose of isolating a section of plant, or a vessel.

6.5 CONTROLLING MAINTENANCE WORKERS

A useful, 'see at a glance' method of keeping in touch with the whereabouts of all maintenance workers, including contractors, in a process area is to use a display system which consists of a large scale plan on a steel backing, together with magnetic flags which identify the various types of activity and operative. This plan should be displayed in the plant control room or similar central location.

6.6 MODIFICATIONS

The carrying out of alterations and modifications to a process plant without the work being properly controlled or recorded, has been the main cause of a number of serious incidents.

There are a number of aspects of this formalized control. The meaning of 'modification' needs to be clearly defined. For example, it could include alterations to the 'hardware', eg pipework, alarms and overload settings; or the less tangible changes to control philosophy or to process operating parameters: the 'software'.

A precise procedure for submitting, sanctioning, and recording modifications which have been properly designed and engineered must be laid down and enforced rigidly without exceptions. Proposals for modifications must be vetted so as to ensure that they are seen by all the staff who are capable of identifying the chance that an unforeseen hazard might be introduced. Minor alterations to a low hazard plant could be approved by local management, but those on a potentially dangerous plant must be vetted by specially selected staff.

All modifications which are carried out must result in an updating of the P&I diagram and also any other relevant plant documentation, eg process operating instructions and training manuals. One means of doing this is to give a nominated person the responsibility for ensuring that details of all modifications are passed to the drawing office and that the master drawing and the plant copy are amended whenever modifications are carried out.

Procedures should be established for checking the integrity of control systems and computer software after modifications have been done.

7. REPORTING ACCIDENTS AND DANGEROUS OCCURRENCES

Statutory regulations[30] require that an employer notifies the HSE (or in some cases other regulatory authorities) of accidents, industrial diseases and dangerous occurrences. The regulations are complex and detailed and should be referred to to establish precisely what sorts of incidents should be reported. Guidance may be obtained from an HSE publication[31].

8. PERSONAL PROTECTION OF THE WORKER

8.1 GENERAL

The basic philosophy of protecting a worker against dangers related to his work activity is to apply safety measures to the dangerous process, rather than to the worker himself. When carrying out plant maintenance work, however, it may upon occasion be unavoidable that the protection has to be applied to the person, perhaps for only a relatively short period of time.

Typically, in the industries under consideration in this book, personal protection would include protective clothing, hard hats, face visors and goggles, and respirators and breathing apparatus.

8.2 LEGAL

For certain work activities personal protection is required by law to be supplied and kept in good condition by the employer, and it is the duty of the employed person to use it and to take reasonable care of it.

The following is a brief guide to the legal requirements regarding personal protection, but only for the types of work relevant to this book. For detailed information the appropriate regulations should be referred to.

PROTECTION OF EYES REGULATIONS 1974
These specify processes for which approved eye protection is required.

ELECTRICITY AT WORK REGULATIONS 1989
For work on or near unprotected electrical apparatus, suitable protective equipment is required.

FACTORIES ACT 1961, SECTION 30
For work in confined spaces, belts and ropes are required and, where dangerous fumes are likely to be present, oxygen and reviving apparatus.

CONSTRUCTION (WORKING PLACES) REGULATIONS 1966
Where there is a risk of falling, safety nets or belts should be made available if secure hand and foot holds cannot be provided.

CONTROL OF LEAD AT WORK REGULATIONS 1980
When persons may be exposed to lead, protective clothing and respiratory protection should be provided.

CONSTRUCTION (GENERAL PROVISIONS) REGULATIONS 1961
In a variety of construction operations, the following may be required: eye protection, respirators, life-saving apparatus, and weatherproof clothing.

CONTROL OF ASBESTOS AT WORK REGULATIONS 1987
ASBESTOS (LICENSING) REGULATIONS 1983
ASBESTOS (PROHIBITIONS) REGULATIONS 1983
When personnel may be exposed to asbestos, the requirements of the above regulations for personal protective equipment are extensive and detailed[25, 26, 27].

IONIZING RADIATIONS REGULATIONS 1985
To give protection against ionizing radiations, a variety of personal protective equipment is required; also the wearing of film badges.

8.3   EYE PROTECTION

Eye protection will be required for such operations as chipping, knocking out bolts, nuts, or similar articles from a structure and the maintenance, dismantling or demolition of plant which may contain acids, alkalis, or other dangerous corrosive materials. Screens, goggles and face visors are the preferred type of protection.

Equipment is available which gives eyes specific protection against burns, lacerations, contusions or radiation. The glasses or goggles which are provided must match the hazard: all-purpose goggles do not exist. In some cases eye protection may have to be worn for long periods, so a properly fitting frame which is comfortable for the wearer should be provided.

Full details of legal requirements are given in the Protection of Eyes Regulations 1974 which include the issuing, availability, replacement, construction and marking of protective equipment; and also the duties of employed persons in using the protection provided. British Standard 2092[32] gives further information.

The following advice is relevant to the use of contact lenses. They may present an additional hazard when a person is in an environment where he may be may be exposed to chemicals, infra-red radiation, contact with foreign bodies, and radiation from welding. Accidental displacement of them may cause dangerous loss of vision temporarily. Thus the use of contact lenses should be prohibited in certain situations and both workers and supervisors should be aware of this. It is not feasible to remove contact lenses and to replace them with prescription safety spectacles, as it takes at least 24 hours for the eyes to adjust.

8.4 RESPIRATORY PROTECTION

Respiratory protection must be used when the air in a working area is not fit to breathe due to the presence of toxic or harmful gas, vapour or dust, or is deficient in oxygen. All these hazards have been dealt with fully earlier in this booklet. Respiratory protection should, in the main, only be used for relatively short term maintenance tasks; it should not (unless it is unavoidable) be employed for regular activities on a plant or for lengthy maintenance work, when an adequate ventilation system should be provided.

Canister (or filter) respirators may be used if there is a free flow of fresh air, which is not deficient in oxygen and from which only small quantities of dust or fumes are required to be removed. Filters specific to the hazard in question should be used as there is no such thing as a universal filter. See BS 2091[33].

Self-contained or air line breathing apparatus must be used if there is the likelihood that the air may contain large quantities of a contaminant, or if there is oxygen deficiency. Breathing apparatus should only be used by specially trained persons. It is a legal requirement that it is checked and inspected regularly. Certification of the testing of breathing apparatus is covered by an HSE memorandum[34]. After use, respiratory protective equipment should be serviced, cleaned, disinfected and then stored in a clean and secure place until it is next required.

A full and very useful treatment of the subject of respiratory protection is given in Carson and Mumford[35]; other advice is given by the HSE[36].

8.5 PROTECTIVE CLOTHING

Under this heading will be considered not only protective clothing, but also associated matters such as equipment for the protection of the ears, head, hands and feet.

In order to provide a safe system of work during plant maintenance, the use of protective clothing may have to be employed to safeguard the operative. Such clothing must be capable of providing the protection required, whilst allowing work to be performed satisfactorily and comfortably. It must also, of course, be actually worn when necessary. To help meet this requirement, the equipment must be inoffensive to wear, so that it is worn as a matter of course, and rules about its wearing must be strictly enforced. It is obviously preferable that people use the protective clothing of their own free will; it will help if they are able to be involved in its choice, and understand the hazard and the need for protection against it. It may, however, be necessary to remind persons of their duty under Section 7 of the HSW Act to take reasonable care of themselves and others at work. This means, in effect, that if the employer specifies protective clothing as part of a safe system of work, then the employee must wear it.

Protective clothing should not be worn outside the plant area, eg in dining or mess rooms.

Detailed information on the type of protective clothing appropriate to a particular hazard can be found in various manufacturers' catalogues and in Carson and Mumford[35].

8.6   HARNESS, SAFETY BELTS AND LIFELINES

The half harness is used in situations where a horizontal pull for rescue is required, eg pulling a man out of a storage tank. The full harness is used where a vertical pull is required, eg rescuing a man from a vertical column. A safety belt is used to prevent an operative from falling where no secure foothold or handhold is available.

All these are subject to periodic statutory examinations and they should conform to the appropriate British Standard.

Lifelines are wire-cored ropes, usually in 20 metre lengths, which are attached to the harnesses. They should be properly stored, regularly inspected and replaced if they show any signs of damage or deterioration.

9. THE STATUTORY EXAMINATION OF PLANT AND EQUIPMENT

It is a legal requirement, particularly in the Factories Act 1961, that certain items of industrial plant and equipment are subjected to regular examination and testing. For the convenience of those organizing maintenance programmes, details of the statutory examinations which are relevant to the subject matter of this booklet are given below.

OPERATIONS AT UNFENCED MACHINERY REGULATIONS 1938
Transmission belts used in the continuous processes scheduled under the regulations.

CONSTRUCTION (GENERAL PROVISIONS) REGULATIONS 1961
Excavations, shafts, tunnels, etc at construction sites and the timbering thereof.

CONSTRUCTION (WORKING PLACES) REGULATIONS 1966
Scaffolds used for building operations.

CONSTRUCTION (LIFTING OPERATIONS) REGULATIONS 1961
Automatic safe load indicators on cranes used in construction operations.
Hoists and lifts used in construction operations.

FACTORIES ACT 1961
- Sections 22 and 25 — hoists and lifts.
- Section 26 — certain chains, ropes and lifting tackle.
- Section 27 — cranes and other lifting machines at factories and on construction sites.
- Section 30 — breathing and reviving apparatus and belts and ropes used for work in confined spaces.
- Section 35 — steam receivers
- Section 36 — air receivers

EXAMINATION OF STEAM BOILERS REGULATIONS 1964
Steam boilers

Note: The requirements of the three preceding pieces of legislation (for steam receivers, air receivers and steam boilers) will be replaced on 1 July 1994 by those of the Pressure Systems and Transportable Gas Container Regulations 1989, to which references should be made for details of periodic examinations after that date.

IONIZING RADIATIONS REGULATIONS 1985
Dosemeters, dose rate meters and sealed sources of radioactive substances.

DOCKS REGULATIONS 1988
Derricks, other lifting machinery, wires, ropes, and lifting tackle used in the loading and unloading of ships, etc.

FIRE PRECAUTIONS ACT 1971
Fire alarms

APPENDIX — EXAMPLES OF ACCIDENTS RELATED TO PLANT MAINTENANCE

1.   ISOLATION OF PLANT

1.1

A serious fire occurred on a distillation plant, in which three men were killed, one seriously injured, and the plant extensively damaged. Maintenance staff had been working on a pump and had decided to dismantle it. When they removed the cover from the pump, hot oil which was above its auto-ignition temperature was ejected and self-ignited, because the pump suction valve had not been closed to isolate the pump from the supply of oil before dismantling commenced. The use of a valve for isolation is not correct; the pump should have been blanked-off from the supply.

1.2

A fatal accident occurred when a process operator accidentally opened the suction valve on an ammonia pump which was undergoing maintenance and was partially dismantled, instead of opening the suction valve on the spare pump. He was killed by escaping ammonia. The accident would have been prevented if the pipework to the pump under maintenance had been spaded-off.

1.3

An explosion in a reactor killed two men. The reactor had been cleaned out and gas-freed before carrying out a maintenance operation. Since there was no intention either to carry out hot-work or to enter the vessel, it was decided not to blank it off but to rely on closed valves for isolation. Some flammable vapour did, in fact, leak into the vessel via a valve and this was ignited by a grinding wheel and exploded.

2.   COMMUNICATIONS

2.1

A man was seriously injured by acid which was ejected from a pump on which he was working. The pump had been isolated by the use of blanking plates, but these had been put in the wrong place and so were not effective. The permit-to-

58

work had properly required that the pump should be isolated but did not specify precisely where the blanks should be inserted.

2.2

A fitter was to repair a leaking joint in some pipework carrying water, on a pipe bridge. Staging was erected but, because of the difficulty of access, the process supervisor pointed out the joint to the maintenance supervisor from the ground. The maintenance supervisor, in turn, pointed it out to the fitter. The fitter opened a joint in a carbon monoxide line in error, was gassed, and fell to his death from the staging. Had the joint to be repaired been identified by affixing a tag, the fitter would not have opened the wrong one.

3. ENTERING CONFINED SPACES

3.1

Because the atmosphere in a vessel was rather unpleasant, a man connected up a hose to what he thought was a compressed air line and put the other end into the vessel. The line was actually from a nitrogen supply. Fortunately the mistake was spotted before he was asphyxiated.

3.2

The normal procedure for entering a tank was omitted because it had contained only water and was not connected to any other equipment. Unfortunately, over the long period of time during which the tank had been left sealed up, rusting had reduced the oxygen content of the air inside it to a dangerously low level. Three men who entered the tank were overcome and one of them died.

4. HOT-WORK

4.1

Repairs had to be carried out on the roof of a storage tank which had contained heavy oil. The tank had been cleaned out as thoroughly as possible and two welders started work outside on the roof. They noticed smoke coming out from the vent pipe and flames from the hole which they had made. They started to leave the area, but before they could do so the tank blew up and an 80 feet long flame was emitted. One of the men was killed and the other seriously burned. The tank had been cleaned out and freed from oil as thoroughly as possible, but traces of heavy residues were trapped between the plates, behind rust or had adhered to the sides. The heat from the hot-work thermally decomposed these residues and ignited the flammable gases which were produced.

4.2

One of the best-known incidents involving hot-work and heavy residues occurred at Dudgeon's Wharf in London in 1969. A tank which had contained a liquid resin was undergoing demolition. Despite having been emptied and steamed out, a gummy deposit remained on the walls and roof. When a flame-cutting torch was applied to the outside of the tank, the heat decomposed the residue to produce flammable gases. These exploded and six men were killed.

4.3

A permit-to-work was issued for the removal of a blanking plate from a 12 inch diameter pipe which contained naphtha — a highly flammable liquid — which was located in a pipe trench. When this was done, several gallons of naphtha leaked out, evaporated and the vapour was ignited by a welding operation some 65 feet away. One of the men who was working on the blanking plate was killed. The vapour from a small leakage of a low flashpoint liquid would not normally spread as far as 65 feet and still form a flammable mixture with air but, in this case, the pipe trench was flooded with rainwater and the liquid spread over the surface to the welding operation. Clearly the permit-to-work was inadequate, the originator having not foreseen the leakage of naphtha when the joint was opened and the possibility of the flammable liquid/vapour spreading so far.

The incident was due, in part, to the fact that it took place some 500 yards from the process plant and that the supervisors there felt that their primary responsibility was for plant operation, rather than for maintenance work. For this reason no-one visited the scene before issuing the permit for removing the blanking plate, but it had been visited for the welding permit. Thus the significance of the flooded pipe trench was not noticed.

5. MODIFICATIONS

5.1

A special system of pipework was installed for compressed air which was to be used with breathing apparatus only, a dedicated branch being taken off the top of the principal compressed air main as it entered the plant. The system was used for 30 years without any problems. Then one day a man who was wearing a face mask whilst working inside a vessel received a face-full of water which nearly drowned him. Fortunately he was able to signal for help and was rescued. It was found upon investigation that the compressed air main had been renewed and that the branch to the plant had been repositioned at the bottom of the supply main rather than at the top. When a quantity of water entered the system, it drained into the breathing apparatus main. Because the system had first been installed so long ago, no-one appreciated the reason for the branch being on top of the main.

5.2

A similar incident occurred on a fuel gas system. When a corroded main was renewed, a branch to a furnace was taken off the bottom of the main instead of from the top. A slug of liquid in the gas filled up the catch-pot, entered the pipework which supplied gas to the burners and extinguished the flames. Upon re-ignition from the hot refractory, there was a serious explosion.

6. EQUIPMENT SENT OFF SITE

A large shell and tube heat exchanger was sent to a workshop for retubing. The tubes contained traces of process material. They were first cleaned out by steaming, except for certain of them which were completely blocked and so steam could not be admitted into them. After most of the tubes had been removed, some men entered the shell to grind out the remaining ones. When this was being done, toxic fumes were produced from the remaining process material, which seriously affected the men.

The maintenance department had asked the plant management if it was safe for men to enter the shell and was told, correctly, that the shell side of the exchanger was clean, but not that some hazardous material could still be inside the tubes.

REFERENCES

1. Health and Safety Executive, *Memorandum of guidance on the Electricity at Work Regulations 1989*, HSE Guidance Booklet HS(R)25, HMSO, London.
2. BS 6626, Code of practice for maintenance of electrical switchgear and control gear for voltages above 650 v and up to and including 36 kv.
3. BS 6463, as reference 2, for voltages up to and including 650 v.
4. BS 1011, Maintenance of electric motor control gear.
5. *Handbook of industrial loss prevention*, 1967, McGraw Hill.
6. Health and Safety Executive, *General access scaffolds*, HSE Summary Sheet SS3, HMSO, London.
7. BS 1129, Specification for portable timber ladders, steps, trestles and light stagings.
8. BS 2037, as reference 7, but for aluminium.
9. Health and Safety Executive, *Safe use of ladders*, HSE Leaflet IAC/L44, HMSO, London.
10. Health and Safety Executive, *Safety in roofwork*, HSE Summary Sheet SS4, HMSO, London.
11. BS 2830, Specification for suspended safety chairs and cradles for use in the construction industry.
12. Health and Safety Executive, *Suspended cradles and small lifting appliances*, HSE Summary Sheet SS5, HMSO, London.
13. Health and Safety Executive, *Industrial use of flammable gas detectors*, HSE Guidance Note CS1, HMSO, London.
14. Mackinson, E.W. et al., 1980, *Pocket guide to chemical hazards*, US Departments of Health and Labor, NIOSH.
15. Health and Safety Executive, *Methods for the determination of hazardous substances*, HSE MDHS Series, HMSO, London.
16. Health and Safety Executive, *Hot work: welding and cutting on plant containing flammable materials*, HSE Guidance Booklet HS(G)5, HMSO, London.
17. Health and Safety Executive, *Entry into confined spaces,* HSE Guidance Note GS5, HMSO, London.
18. BS 5980, Code of practice for site investigations.
19. BS 6031, Code of practice for earthworks.
20. BS 8004, Code of practice for foundations.
21. *Safe handling and storage of compressed gases: gas data and safety sheets*, BOC Special Gases Ltd, 1980.

22. Health and Safety Executive, *Keeping of LPG in cylinders and similar containers*, HSE Guidance Note CS4, HMSO, London.
23. *Safety in chemical engineering research and development*, 1991, Institution of Chemical Engineers, Rugby, UK.
24. Health and Safety Executive, *Occupational exposure limits*, HSE Guidance Note EH 40, HMSO, London.
25. Health and Safety Executive, *Provision, use and maintenance of hygiene facilities for work with asbestos insulation and coatings*, HSE Guidance Note EH 47, HMSO, London.
26. Health and Safety Executive, *Removal techniques and associated waste handling for asbestos insulation, coatings and insulating board*, HSE Guidance Note EH42, HMSO, London.
27. Health and Safety Executive, *Probable asbestos dust concentrations at construction processes*, HSE Guidance Note EH 35, HMSO, London.
28. Health and Safety Executive, *Guide to the Asbestos Licensing Regulations 1983* HSE Regulations Booklet HS(R)19, HMSO, London.
29. Health and Safety Executive, *Safety in the use of cartridge operated tools*, HSE Guidance Note PM14, HMSO, London.
30. *Reporting of Injuries, Diseases and Dangerous Occurrences Regulations 1984*, (RIDDOR).
31. Health and Safety Executive, *Guide to RIDDOR*, HSE Regulations Booklet HS(R)23, HMSO, London.
32. BS 2092, Specification for eye protectors for industrial and non-industrial uses.
33. BS 2091, Specification for respirators for protection against harmful dusts, gases and scheduled agricultural chemicals.
34. Health and Safety Executive, *Certification of breathing apparatus*, HSE Testing Memorandum TM3, HMSO, London.
35. Carson, P.A. and Mumford, C.J., 1988, *The safe handling of chemicals in industry*, Longman Scientific and Technical, Harlow.
36. Health and Safety Executive, *Respiratory protective equipment: a practical guide for users*, HSE Guidance Booklet HS(G)53, HMSO, London.

Note: The Health and Safety Executive publishes a wide range of guidance literature, much of which is relevant to the subject matter of this book.

A list entitled 'Publications in series. List of HSC/E publications' is obtainable from:

The Health and Safety Executive
Information Centre
Broad Lane
Sheffield S3 7HQ

INDEX